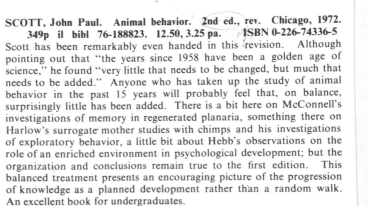

SCOTT, John Paul. Animal behavior. 2nd ed., rev. Chicago, 1972.
349p il bibl 76-188823. 12.50, 3.25 pa. ISBN 0-226-74336-5
Scott has been remarkably even handed in this revision. Although
pointing out that "the years since 1958 have been a golden age of
science," he found "very little that needs to be changed, but much that
needs to be added." Anyone who has taken up the study of animal
behavior in the past 15 years will probably feel that, on balance,
surprisingly little has been added. There is a bit here on McConnell's
investigations of memory in regenerated planaria, something there on
Harlow's surrogate mother studies with chimps and his investigations
of exploratory behavior, a little bit about Hebb's observations on the
role of an enriched environment in psychological development; but the
organization and conclusions remain true to the first edition. This
balanced treatment presents an encouraging picture of the progression
of knowledge as a planned development rather than a random walk.
An excellent book for undergraduates.

Animal Behavior

John Paul Scott

Animal Behavior

Second Edition, Revised

The University of Chicago Press

CHICAGO & LONDON

The University of Chicago Press, Chicago 60637
The University of Chicago Press, Ltd., London
© 1958, 1972 by The University of Chicago
All rights reserved. Published 1972
Printed in the United States of America
International Standard Book Number: 0-226-74336-5 (clothbound)
Library of Congress Catalog Card Number: 76-188823

Contents

Preface

It has now been fourteen years since the first edition of *Animal Behavior* appeared. As I reread it I find very little that needs to be changed, but much that needs to be added. The years since 1958 have been a golden age of science in the United States, thanks to the generous support of the federal government, and this has been reflected in the science of animal behavior. Whereas in the 1950s there may have been a dozen or two active workers in the field, there are now hundreds. Where formerly the important communications in animal behavior could be reported in a single combined session of the American Society of Zoologists and the Ecological Society, the Animal Behavior Society now organizes two crowded programs a year, each extending over three days. Similar advances have occurred in other countries where financial support of science has been increased. The once small and intimate Ethological Conferences of Europe now attract hundreds of participants. The general result is that where formerly an important scientific result might be reported once every five or ten years in a field like animal behavior, such results have recently been appearing every year or two. I have tried to include at least the major new advances in this second edition, although it has never been the purpose of this book to include all that has been done. Such encyclopedic treatments are found in other volumes, which also represent major accomplishments of the last few years. Such books as those of Marler and Hamilton and of Hinde are mines of detailed information.

What are some of the special areas of progress? I immediately think of the field studies of primate social behavior, some of which were undertaken by biologists such as Emlen, Southwick, and Schaller, but whose principal inspiration came from a group of physical anthropologists—Washburn, DeVore, and Phyllis Jay among others. Creatively brilliant studies have been done by European scientists such as Jane Goodall and Hans Kummer as well as by Japanese workers who have largely concentrated on the Japanese macaque. Corresponding field studies of other large mammals have also been done, but few of them approach the thoroughness of the primate studies, except perhaps those with wolves.

Another area in which enormous progress has been made is that of animal communication and animal orientation. On the one hand, experimenters have become fascinated by the echolocation and sound production of undersea animals, and on the other the technology of sound reproduction and analysis has made possible an enormous extension of information regarding the sounds of land-living animals. Insects in particular have been studied by the group of French scientists led by Busnel. Equally interesting are the advances in understanding communication through chemical signals, the so-called pheromones. These odorous substances transmit important and extensive information in insects, and more recently their function has begun to be appreciated in mammals.

As the first edition stated, animal behavior can be studied from the viewpoint of many levels of organization and their corresponding scientific disciplines, and a complete and reasonably accurate picture of the causes of behavior can be reached only by taking all of these into account. As progress has been made in analyzing the genetic causes of behavior and the physiological mechanisms through which they work, many of the old and rather sterile disputes over the importance of genetic and environmental factors have begun to disappear. Each is essential, and each includes many different factors. For further progress we need not only an interdisciplinary approach, but also conceptual and mathematical tools by which we can measure and predict the effects of many factors co-acting and interacting with each other. Unaided by these, most thinkers are unable to get beyond the simple dichotomous thinking represented by the heredity-environment controversy.

These new tools are nowhere more badly needed than in the applications of animal behavior to human affairs. The facts of genetics demonstrate that every species is genetically unique. Therefore it is not possible to analogize directly from one species to another. At the most, results from one species suggest only possibilities for another. Drawing broad generalities from animal behavior can only be done from the demonstration that a phenomenon exists in a large and representative sample of the animal kingdom. Even when the phenomena are similar, the underlying mechanisms may be quite different. If animal behaviorists are to speak knowledgeably and authoritatively about the relevance of their science to human affairs (and I believe that they should), they must either become sophisticated concerning the many factors affecting behavior in their own sciences as well as the peculiar factors that are unique to human behavior and form a large part of the subject matter of the social sciences, or they must speak humbly and circumspectly.

The future of the science of animal behavior depends on many factors, not the least of which is the relative amount of time devoted to positive and constructive activity, as opposed to destructive activity, permitted by the society in which we live. Nevertheless I believe that it will go forward, partly because of the hope it affords for achieving a better human social organization, but above all because of its intrinsic interest. Furthermore, a great deal can still be done with no more elaborate tools than a paper and pencil and perhaps binoculars or a stopwatch. The behavior of the vast majority of animal species has still not been explored or analyzed, and if past experience is any guide, each new species studied will reveal new and fascinating facts.

Preface to First Edition

This book has been written to answer the general question, What is the study of animal behavior about? It is also intended as a textbook, which means that it attempts to give a fair and accurate picture of a field of knowledge. It is designed for the general reader or the student who wishes to learn something for himself. With this in mind I have tried to use a minimum of scientific phraseology, so that the book can be read without interruptions or outside help. At the same time I have made no effort to disguise the fact that animal behavior is a genuine science and, as such, must employ clearly defined terms and precise logic in order to produce true general principles.

The subject matter is organized according to the groups of factors affecting behavior which operate at each level of biological organization. Some of these can be expressed as basic laws of behavior and are illustrated with examples selected from the animals upon which important research work has been done. These examples do not systematically cover the animal kingdom, and the teacher who uses this book as a text will probably wish to consider it as an outline that he can expand and amplify in his lectures and through additional reading. The bibliography is arranged as a guide to the latter.

Animal behavior is concerned with the activity of the whole organism and of groups of organisms. Its study therefore involves the use of techniques and principles from all branches of zoology—anatomy, physiology, ecology, genetics, and even embryology and taxonomy. It is also an interdisciplinary study on a larger scale, involving findings from psychology, sociology, and the physical sciences. In past years there has been a tendency to divide zoology into compartments, each with its corps of specialists following specialized and unrelated problems. It is hoped that in this book the student will find a glimpse of the essential unity of animal science, which is not the artificial result of applying an oversimplified theory but a natural unity of ideas arising from the effort of explaining the important problem of what an organism does. This book is written with the assumption that what an organism does is more important than what it is, and that behavior is one of the central problems of existence.

One way of assessing the importance of a subject is to consider it historically. Science does not exist in a vacuum but is a part of human behavior and particularly part of the behavior of the scientists and students who work with it. These are some of the people who have worked on the science of animal behavior, with a brief account of what they have done.

The study of animal behavior has fascinated mankind ever since the times of Solomon and Aesop. By long historical tradition people are willing to learn from the ant or the fox what they refuse to see in their fellows. However, this subject did not receive the serious attention of scientists until the middle of the nineteenth century, when Darwin's theory of evolution placed great stress on the idea of progressively improved adaptation. The adaptation of an animal is, of course, largely accomplished through its behavior, and Darwin himself devoted a great deal of attention to the subject. His books *The Formation of Vegetable Mould through the Action of Worms* and *The Expression of the Emotions in Man and Animals* still provide useful and accurate information.

His lead was followed by a large number of important European and American scientists, and the bibliography of Jennings' *Behavior of the Lower Organisms*, published in 1906, reads like a Who's Who of early twentieth-century scientists. Among American biologists appear the names of T. H. Morgan, Jacques Loeb, Raymond Pearl, E. B. Wilson, G. H. Parker, S. O. Mast, and S. J. Holmes, while the Europeans include such distinguished names as Claparède, Driesch, Lloyd Morgan, Naegeli, Pavlov, Romanes, Verworn, and Von Uexkull.

Meanwhile, the psychologists Yerkes and Thorndike were turning their attention to the study of the behavior of the higher animals, and the results of a fruitful collaboration between psychologists and biologists appeared in the *Journal of Animal Behavior*, in six volumes published before World War I.

Two significant scientific discoveries diverted the attention of these workers into other fields. One was the rediscovery of Mendelian inheritance in 1900, when the majority of biologists, including such workers as Morgan, Pearl, and even Jennings himself, turned their attention to developing the new science of genetics. About the same time, the discovery by Pavlov of the conditioned reflex seemed to provide a basis for rapid progress in the scientific analysis of learning. Psychologists such as Thorndike found that the white rat was a convenient and inexpensive animal for these studies and no longer concerned themselves with the broad problems of adaptation. By the end of World War I, interest in the science of animal behavior seemed almost dead.

Up to this time the problem of adaptation had been approached almost entirely from the viewpoint of individual survival, and the study of social behavior in animals had been so limited to insects that it was scarcely more than a branch of entomology. Around 1920 came two new discoveries in the behavior of birds which indicated that the behavior of other animals was significant in the context of a complex social organization. Howard's fresh insights into the significance of song and territory and Schjelderup-Ebbe's description of social dominance in hens were the first of a series of exciting new discoveries. Shortly afterward, Allee published his first work on animal aggregations, to be followed by a long series of studies on basic animal sociology. C. R. Carpenter made a study of the social relationships and organization of free-living primates, and in 1935 Lorenz published his studies of the formation of primary social relationships in birds. Meanwhile, progress had not been lacking in the study of the social behavior of insects. Wheeler had presented his theory of trophallaxis as the basis of social organization in insects, to be followed by Emerson's extensive studies on social differentiation in the termite, with consideration of the animal society as the unit of evolution. Schneirla studied the complex societies of the army ants, and Von Frisch was able to establish experimentally the existence of a "language" in bees.

World War II interrupted many of these studies, particularly in Europe, but since then there has been a great revival of interest on both continents. Tinbergen and other European workers have become interested in the problem of instinct as it relates to social behavior and have founded the journal *Behaviour*, devoted to the study of comparative ethology, or the comparison and analysis of behavior traits in different species. A number of British ornithologists, including Armstrong, Lack, Thorpe, and others, are exploring the almost infinite variety of social behavior found in birds. A large group of younger American biologists, among whom may be mentioned Calhoun, Collias, Emlen, Nice, Kendeigh, Davis, Guhl, and King, have become interested in the problems of social organization and its relation to population dynamics. Another group, of which the author is a member, have devoted their attention to the study of Mendelian genetics and social behavior. Such workers as Beach and Young have made great contributions to the analysis of the physiology of social behavior. By and large, psychologists are tending to leave these studies in the hands of the biologists, but there is a growing number of notable exceptions, such as Hall, Nissen, Liddell, Harlow, Hebb, and Thompson, as well as an active group devoted to the neurological analysis of behavior.

For the last quarter-century, the impetus and interest in the study of animal behavior has come largely through the general concepts of social organization and behavior. It is the intention of this book to present some of the most important problems which now face this science.

Acknowledgments

Science is a collaborative enterprise, and this has become increasingly apparent to me as I have prepared the two editions of this book. Without the efforts of the many individuals who have dedicated their lives to the study of animal behavior, there would be very little to write about. Whether we like it or not, each of us must link his own work with that of his colleagues and those who have gone before.

In particular I want to thank those individuals who have contributed many of the excellent photographs, and to whom credit is given in the text. The second edition still shows the results of the photographic and editorial help given by Ralph and Mildred Buchsbaum during the preparation of the first edition. I also want to thank my daughter, Vivian Hixson, for preparing many of the drawings and my wife, Sarah F. Scott, for her encouragement, editorial judgment, and objective criticism.

CHAPTER 1

Animal Behavior
and Human Behavior

A familiar autumn sight in many parts of the northern and central United States is the appearance of great flocks of blackbirds which settle down to feed on the stubble fields. As we look out over the pleasant countryside we can almost always see at least one field carpeted with feeding blackbirds. The group is spread out over a hundred feet or so, and we notice that as the birds hop industriously about, turning over leaves and picking up bits of food, they manage to keep about the same distance apart from each other so that the effect is always that of a compact group rather than a random collection of individual birds.

Suddenly a few birds on one edge of the flock become disturbed and fly up. The others next to them lift into the air in succession until the whole flock rises and wheels as a unit as it goes off to repeat the performance in some neighboring field. At dusk blackbird flocks may come into a nearby town and roost on specially favored shade trees, returning night after night to the same spots, much to the distress of the human owners who wish to keep the sidewalks clean.

As we watch we begin to wonder. Why do the birds stay close together, when an individual bird might easily find more food and have it all to himself? Do all the birds belong to the same species? Does each flock have a leader? How can the birds coordinate their behavior so exactly, and why do they prefer certain trees above others? One question leads to another, and finding the answers is the heart of the science of animal behavior.

THE USES OF ANIMAL BEHAVIOR

Why should we study the behavior of animals? The answer given to this question by any particular student is likely to be an indi-

1

vidual one. Some of us might paraphrase Mallory's famous reason for climbing Mount Everest—because it is there—and say, "Because it is unknown." More intensive study of the flocking blackbirds reveals even more remarkable and mysterious activities, for later in the autumn they suddenly vanish, not to reappear until the next spring. In the meantime they may have traveled hundreds or thousands of miles, yet are able to return to the same places the following year. Finding the explanation is a challenge to imagination and ingenuity. All sorts of methods have been tried, from laboratory experiments to pursuit by airplanes, and the problem of bird migration is still not completely solved.

Others might give the more prosaic reason that the behavior of blackbirds has economic consequences. Red-winged blackbirds live on the food crops of man as well as on insects, and their damage in a recent year in northern Ohio was estimated at 10–15 million dollars. On a smaller scale, how much is it worth to the indignant householder to get the blackbirds out of his trees? Certainly the European starlings have become a serious nuisance in American cities, where they apparently like to keep their feet warm by roosting near electric signs.

An equally pressing and concrete reason for studying animal behavior is that for thousands of years people have observed the activities of animals and, rightly or wrongly, applied their conclusions to human nature. A correct application can be extremely helpful, but a wrong one can produce results ranging from comic to disastrous. In the last century, when horses provided the common means of transportation in the United States, certain methods were developed for breaking and training them to harness. Many people applied these stern methods to their own children. It is still possible to find very old people who will say that an ideal child must show absolute obedience and that, in order to train him properly, it is necessary to "break his will." This is all very well for a horse, which is a large and potentially dangerous animal, but would such methods be likely to produce an ideal human citizen with some judgment of his own? Whether we like it or not, people will apply their ideas of animal behavior to human behavior. This being so, it is highly desirable that these ideas be as correct as possible, and this is something which can be achieved through scientific study.

The comparative method. We are so used to the system of

standard units of measure that we are apt to forget that the essential process of any measurement is comparison. If we want to know the height of an animal, we compare it with a piece of wood or metal graduated in standard lengths; or if we want to know its weight, we compare it with pieces of metal which have in turn been compared with weights kept at the Bureau of Standards. In the case of behavior, certain common standard measures can be used, such as time and frequency. However, many aspects of behavior have not been universally standardized, and it is often convenient to use the behavior of one species as a standard of comparison with another. This can be legitimately done between any two species. Indeed, one of the chief techniques of the science of animal behavior is an exact systematic comparison of behavior, checking off likenesses and differences. When the method is applied to the human species, however, certain general principles need to be kept in mind in order to produce significant results.

The use of new ideas. Studying the behavior of any animal inevitably gives rise to new ideas. After watching the precise and mutually coordinated flying of blackbirds, we begin to wonder whether human beings and other mammals do the same sort of thing on the ground. Further observations reveal that the behavior of herd animals like sheep is very similar to that of the blackbirds, and that there are certain situations in which even human behavior may correctly be described as "sheeplike." When an interesting phenomenon is observed in any species, it immediately raises the question: What if this should be so of people? The answer must be given by direct observation and experiment. It is not correct to reason by analogy and say that because a certain statement is true of rats it must be true of human beings. As will be shown in the following pages, many of our greatest new ideas and insights regarding human behavior have come from animal studies, but the final test of their correctness always depends upon the direct study of human beings.

The significance of animal experiments. Psychologists and psychiatrists frequently get ideas about behavior from their experiences with fellow human beings. For example, many clinical experts believe that certain kinds of mental troubles have their origin in the early experience and training of children. It is impractical to test these ideas experimentally on people because of the length of human life and undesirable because, if they are correct, we would not wish to inflict the results upon children. Attempts are therefore made to experiment with various kinds of early experience in young animals, with the hope that this will shed some light on human behavior. As we shall see, many of these experiments have turned out to be inconclusive, but certain kinds of early experience, such as being raised apart from the parent species, produce some remarkable effects on later behavior. We have gained an inkling of what this might mean in terms of human experience, but it is still too early to make direct applications.

It must be added that such experiments are full of pitfalls for the enthusiast. Both the behavior of the experimental animal and the corresponding human activity must be so well known that we can be certain we are working with identical phenomena. Furthermore, we must be sure that the experience means the same thing to the animal as it does to the human being; and, since many animals have widely different sense organs and motor capacities, this may not be the case. A far more important result of such experiments is the eventual formation of generalizations.

General laws of behavior. One of the primary objectives of science is to explain what we see around us by the use of a limited number of broad general ideas. In the case of behavior, these can be obtained only by careful study of a large and representative sample of the animal kingdom. If it is found that a statement is true of a very large proportion of all of the species studied, it is correct to assume that the same statement is very likely true of human beings and that it concerns an important and basic element in human behavior. Amassing the needed information is a tremendous task. Nevertheless, as scientists have observed and analyzed the behavior of hundreds of different animal species, they have developed at least one basic generalization, the stimulus-response theory.

THE STIMULUS-RESPONSE THEORY

Essentially this is a very simple idea: That behavior, which is called a response, always has some sort of cause which precedes it. In mathematical symbols it may be written $S \rightarrow R$. The cause or stimulus may be inside the animal or outside it, it may be a change in the physical or social environment, but it is always there. Consequently the study of stimulation is an important part of the study of behavior.

A stimulus is always a change. One of the groups of animals in which it is easiest to study stimulation is the Protozoa. These tiny one-celled animals can be kept in a single drop of water under a microscope while experiments are done with them. They were studied extensively by H. S. Jennings in the early 1900s, and his book, *The Behavior of Lower Organisms,* is still a standard reference and a model of clear exposition. He found that almost any sort of environmental change will produce a response. A euglena, for example, is a small green organism, familiar to most students in beginning biology. It is shaped like a fish and has, at the anterior end, a red pigment spot and a long flagellum, by which it constantly pulls itself through the water in a spiral course. If it is placed in strong sunlight, it turns and swims away; but if the light is weaker, it swims toward the source. If the light is cut off, the euglena stops, swings in a circle, and then eventually starts out in the general direction of any light which may be left. The same sort of thing happens if the direction of the light is changed. In short, any change in the intensity or direction of light causes the animal to stop, circle, and start off, either toward or away from it.

A constant set of conditions does not act as a stimulus. If a paramecium is dropped into a 0.5 percent solution of ordinary salt, it will at first swim jerkily backward and forward; but in a few minutes it apparently adapts to the new situation and swims ahead in its usual fashion. Similarly, a paramecium may be living normally in water at 20° C. If the temperature is suddenly lowered a few degrees, it will at first give an avoidance reaction; but if the temperature is kept constant, it soon acts just as it did before the change.

The same phenomenon can be seen in higher animals. A young puppy has a tendency to give a startle reaction to any sudden noise. When an electric fan is turned on, he jumps. The noise continues, but he does not continue to react to it and soon pays attention to other things. When new sounds occur, he seldom reacts to them until the fan is turned off, for noise against a background of sound is less of a change than a new noise against a long-continued silence.

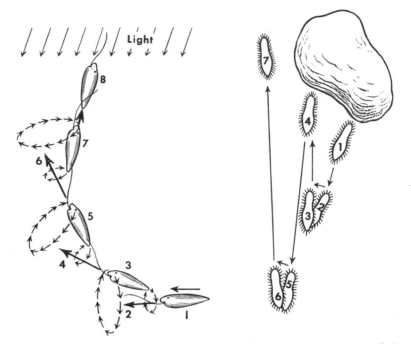

Fig. 1. Stimulus-response theory is illustrated by the behavior of one-celled animals. *Left*, euglena responds to a change in the direction of weak light. As the animal swims in its usual spiral path, the eye spot in the front end swings in a circle. As long as the light comes from directly ahead, the spot receives even illumination. When the light comes from the side, the swinging motion produces a change in each revolution, as the eye spot turns toward and away from the light. Numbers show successive turning responses as the light, which has been coming from the left, comes instead from the top of the diagram. (Modified from Jennings, *Behavior of the Lower Organisms,* 1906, by permission of the Columbia University Press)

Right, avoiding response in paramecium as it adapts to striking an obstacle. The first attempt at *4* is unsuccessful, and the animal backs off for a new try. A stimulus is a change, and a response is an attempt to adjust to change. Both examples also illustrate variability of behavior. (After Buchsbaum, based on Jennings)

In addition, certain of the sense organs of higher animals show the phenomenon of accommodation, which, to some extent, explains the fact that only a sudden change acts as a stimulus. That the human nose rapidly accommodates itself to odors is well known. At first we can smell an odor distinctly, and then it seems to disappear. There is a similar accommodation of sense organs to heat; a person who gets into a hot shower and gradually increases the temperature may actually scald himself without being aware of pain. Even in the case of vision, where ordinarily we are not conscious of accommodation, the eyeball continually makes small movements which ensure that any given cells in the retina never receive light from the same source for more than an instant. If these movements are prevented, vision begins to fade.

The principle that a stimulus is a change has many practical applications in the study of animal and human behavior. Changes in activity occur at times of day when there are large changes in the physical environment. Many species of animals are most active either at dawn or at dusk, having a tendency to be inactive either in the middle of the day or in the middle of the night. A ten-minute observation during the active period will yield more information than an hour when the animal is resting under a bush.

The principle is even more practical when we attempt to stimulate and control the behavior of people. We find that the first statement of good advice may have some effect, but constant repetition soon has no more result than the noise of an electric fan upon the behavior of a puppy. A varied diet is much more stimulating to the appetite than a monotonous one. A practiced speaker varies his tones of voice and arouses his audience, while the droning lecturer who neglects this rule produces only a drowsing row of students. Other examples can be found in the arts of communication and entertainment or in almost any field of human relations.

A response is an attempt to adapt to a change. The green coloring in a euglena is chlorophyll. In the presence of light this animal is able to manufacture its own food by photosynthesis, and most of its behavior is related to this fact. In the fresh-water pools in which it normally lives, its continual reactions to changing light conditions keep it moving toward the moderate sunlight which permits photosynthesis but is still not strong enough to kill the animal. Thus we can say that its behavior is adaptive to changing light conditions.

Not all adaptation is successful. When we put a paramecium in a weak salt solution, it responds by moving more rapidly in various directions. If it were in a large body of water, and the salt were found in only one spot, this behavior would enable it to escape—a successful adaptation. As we set up the experiment there is only one drop of water, all of which has salt in it, and the attempt to adapt is unsuccessful. If it were not that attempts at adaptation so frequently fail, we could say that behavior *is* adaptation. As it is, we can define a response only as an *attempt* to adapt.

The nature of adaptation. The law of adaptation is a basic biological principle which may be stated simply thus: An organism tends to react in ways which are favorable to its existence.

Adaptation is more than a simple reaction to forces. An animal does not behave toward stimulation in accordance with the simple mechanical principles of physics. If a billiard ball is struck a straight blow with a cue, it moves in the same direction as the force applied to it. If you poke a dog with a billiard cue, he may do a variety of things, one of which is to turn around and bite the cue. He may even run around and bite you on the leg, leading to considerable confusion in the experiment. Or he may run off and bark at a distance considerably greater than that to which you pushed him with the cue. In no case does he react like the billiard ball, the motion of which is the resultant of all the forces which act upon it. An organism is not simply pushed around by a number of forces. It attempts to adjust and adapt to change, and the amount of physical force exerted may have very little to do with the kind of reaction that is produced. A deer which hears a snapping twig is stimulated only by a minute amount of energy, but it may leap and run for a quarter of a mile as a result.

Neither does an animal react to all the forces which act upon it. A paramecium which is subjected to a higher temperature constantly receives energy from this source but may give a behavioral response only to the initial change. Likewise, some changes in the environment will be totally disregarded. There is a tendency for an organism to respond to only one stimulus at a time, no matter how many may be occurring in the environment.

Finally, not all behavior is directly adaptive to external and internal stimuli. Like life itself, behavior is a complex, ongoing process that does not cease when adaptations to stimuli are no longer required. The result may be "surplus" activity, largely irrelevant to environmental conditions, which we usually label "play." Again, in the complexly organized higher animals, an individual may be capable of transforming stimuli into something quite different from those originally received, and so in effect create new stimuli. This in turn makes possible novel solutions to problems of adaptation. Once such creative capacities were developed, natural selection must have favored the individuals who possessed them. We are most familiar with this process as creative imagination in human beings, but there is evidence that this capacity may be shared at least to some extent by our nonhuman relatives.

The study of stimulation. The stimulus-response theory suggests a way of dividing up the study of behavior. The idea of stimulation leads to a study of the *causes* of behavior. External stimuli enter the body through the sense organs, and we immediately find that animals differ widely in their sensory capacities. The euglena has a light-sensitive spot, but there is no evidence that it can detect changes in the color or form of objects. This brings up the problem of differential capacities and their effects on behavior. When we attempt to follow the pathways of nervous stimulation still farther within the body, we discover that they are connected to a whole network of internal physiological reactions, some of which may themselves act as stimuli. This, in turn, leads to a study of the function of the nervous system and a consideration of the phenomenon of learning, which forms the central core of the science of psychology. Learning is affected by heredity, as are the sense and motor organs, and this leads to a study of heredity and its effects on behavior. The first part of this book is thus concerned with the *causes* of behavior: the various kinds of external and internal stimulation, and the modification of behavior by learning and heredity.

The study of adaptation. On the other hand, we can concentrate our attention on effects of behavior. How does behavior promote survival? How is it organized to meet the complex problems of daily living? The behavior of one individual affects that of others, and these interactions lead to social organization, a special part of which is the process of communication. Social organization in turn affects the growth and decay of populations. As a final question we may ask: What happens to populations and where do they seem to be going? The latter part of this book is concerned with the *effects* of behavior upon the survival of individuals, the organization of animal societies, and the growth and changes of populations.

To understand either the causes or effects of behavior, it is first essential to establish the basic facts. We must know what behavior is before we analyze its origins and results, and that is where the next chapter begins.

SUMMARY

Like any science, animal behavior may be studied for the simple pleasure of discovering the world around us; but it is so closely connected with the science of human behavior that there are many other pressing and concrete reasons for its advancement. Animal behavior can be a yardstick for human behavior. It gives rise to new ideas which can be tested on human beings, and animals in turn can be used as testing material for ideas derived from human behavior. However, the most important objective of the science is to develop general ideas and theories which explain the behavior of all animals, including ourselves.

One of the most basic and universal behavior theories is that a stimulus causes a response. A stimulus is always a change, and a response is an attempt to adapt to the change. Almost all behavior can be analyzed in these simple terms, and this forms a foundation for the more complex analysis of the causes and effects of behavior. Before this can be done, we need to gather the basic facts about the characteristic behavior of any species studied.

CHAPTER 2

The Elements of Behavior
Methods of Study

In studying the behavior of any animal we first concentrate our attention on the kind of responses it gives to its environment. To state this as a scientific principle, the primary phenomenon to be studied in the science of animal behavior is behavior itself. The most important thing about an animal is *what it does;* and when we have described this, it is possible to go ahead and analyze the environmental changes which cause it to act, the structural and physiological peculiarities which modify its behavior, and the various social and individual consequences of its reactions.

This is not as easy as it sounds. I once had the opportunity of watching the behavior of a herd of wild buffalo in Yellowstone Park, in company with a young scientist who was making a detailed study of their behavior. We were able to approach within half a mile as we climbed to the top of a ridge where we could watch them through a telescope. We could see three or four hundred of the huge animals, grazing and resting in a mountain valley. Calves followed their mothers, and one of the bulls lumbered around and threatened others, who moved out of the way when they saw him coming. Off on the edge of the herd, a bull attacked a young pine tree and tore it to pieces with his horns. Finally the whole herd stirred restlessly and went off down the valley.

What were the buffalo doing? According to the stimulus-response theory, behavior is essentially an attempt to adapt to change. Yet all that I could see in a short observation period were several confusing activities whose causes and adaptive effects were not immediately apparent. The other scientist, who had watched the buffalo over and over again, saw much more than I did, and he could predict approximately what they would do.

What the buffalo were doing becomes clearer when we compare it with the behavior of another species, such as the red-winged blackbirds we have already seen wheeling over the autumn fields. When we compare the activities of a wider variety of species, we begin to see that certain kinds of behavior occur over and over again, and that these fall into a few general kinds of behavioral adaptations which are widely found in the animal kingdom. In making these comparisons we must meet a problem of language as well as that of careful description. What words can we use to say that two animals as unlike as elephants and spiders are doing similar things?

The common terms which are used to describe animal behavior have frequently been developed on the same principles as the terms employed by the medieval experts on hunting, who had a special term for each animal. They spoke of a "covey" of partridge, a "gaggle" of geese, and a "pride" of lions. As we can easily see, the general term "group" will cover all these terms and make comparisons and general statements much easier. In describing animal behavior, some common terms, such as the word "group," are both general and accurate enough to be used directly, but some, like the medieval terms, are too specific to be useful when we wish to speak of the whole animal kingdom. For example, we may speak of "attentive" behavior as blackbirds attend the nest and feed the young, or "maternal" behavior as the buffalo cows allow their calves to nurse, but these behaviors obviously belong to one general type. We need a set of terms which apply to the important elements of behavior of any animal, and with these we shall be able to make comparisons and state general laws.

The behavioral cycle of the red-winged blackbird. A number of years ago Professor Allen of Cornell University made a study of the behavior of the red-winged blackbirds which inhabit the cattail swamps in the region of Ithaca, New York. He found that the birds first appear late in the winter. Flocks of males fly into the swamps at sunset and out the next morning, apparently migrating to the north. As the last ones arrive, the flocks break up, and certain males stay in the swamps permanently. Each of the resident males settles in a separate part of the marsh and begins to drive away the others. Then flocks of migrating females and immature males begin to come into the swamp, flying out again the morning after their arrival. At first the resident males chase them away whenever they come near, but some females finally settle down in the swamp, each with one of the males. The male sits on top of a particular reed and sings while the female gathers food. Courtship behavior is followed by mating. Next the females begin to build nests, and to lay eggs and incubate them. When the eggs are hatched, both parents feed the young and clean the nest, although the female does about three times as much work as the male. When the parents light on the nest, the young hold up gaping bills toward them, and the adults put in the insects which they have collected. Later in the spring the young birds leave the nest, and still later all the blackbirds seem to disappear. However, they may be found hiding deep in the swamps while they molt. In the autumn the sexes separate again to form great flocks in the adjacent grain fields. Later in the fall they fly south to spend the winter.

It is obvious that blackbirds do different things at different seasons of the year. Closer analysis shows that these seasonal activities consist of behavior patterns, each having a specific function and usually organized with others into a system having a common general function.

BEHAVIOR PATTERNS: THE UNITS OF BEHAVIOR

A behavior pattern may be defined as a segment of behavior which has a specific adaptive function. Among blackbirds the behavior of nestlings in rearing their heads and showing the gaping reaction is a behavior pattern, as is the behavior of the parent bird as one thrusts its beak down the throat of the nestling, releases the food, and gives it a downward thrust with its beak. Specific functions of a behavior pattern are not always immediately obvious to the human observer, and it may take a great deal of careful research to determine that function, as it was in the case of bird song to be described in a later chapter. Any regularly repeated sequence of behavior newly observed in an animal may be assumed to be a behavior pattern and a search made for its function.

In order to test whether we are dealing with a single behavior pattern or a series of such patterns, we have only to subdivide the behavior further and see if it retains its specific function. The behavior pattern of the blackbird feeding its young, for example, starts out by its lowering its head. This has no specific function and could be a part of any number of other behavior patterns such as lowering the head while picking up food, or even while building a nest. What we are dealing with here is *movement*, which leads to another kind of behavioral analysis. Movements may concern parts of the body, as in the above example, or movements of the body as a whole, in which case they are called *locomotion*. Understanding how an animal moves is part of the analysis of its behavior, and of course has consequences for what the animal does. Species differ a great deal in their general motor capacities, and this is the subject of a later chapter.

Movement and locomotion, however, give only a very superficial and general picture of behavior. The behavior pattern, which is a combination of movements and often includes locomotion also, provides a more complex and precise way of analyzing behavior. Because a behavior pattern must have a special adaptive function, it forms the behavioral raw material on which natural selection must act. If behavior is not adaptive, it will have no effect on survival. In this sense, then, the behavior pattern is a natural unit.

Systems of behavior. In many cases animals have alternate behavior patterns, each of which provides adaptation in a slightly different way. The male blackbird in the nesting season may either warn other males away by singing or he may actually fly at them and attack them if they come too close. We can therefore say that behavior patterns are organized into behavioral systems. Such a system can be defined as a group of behavior patterns having a common general function.

How can we identify a behavioral system? The first task is to describe the behavior patterns and see if there is any sort of order among them. In the example of the blackbird, it is obvious that these animals do different things at different times of the year, and if certain groups of behavior patterns only occur at certain seasons, we have good empirical evidence that they belong to different systems. Comparing such groups in the behavior of blackbirds with the behavior of other species gives us a list of nine important general behavioral systems. This does not imply that all systems will be found in every animal species; indeed, there are many species in which both behavior patterns and be-

Fig. 2. Systems of behavior, as seen in the red-winged blackbird. (a) Ingestive behavior. The blackbirds gather in large flocks in the autumn and feed on grain and seeds that have fallen to the ground, using their beaks to pick up the food. They also eat unharvested grain when available. (b) Shelter-seeking. At night the autumn flocks of blackbirds seek shelter together, often in the cattail swamps used for breeding grounds earlier in the year. (c) Agonistic behavior. Two males fighting in a dispute over territory early in the breeding season. The two birds peck at each other, but seldom inflict serious damage. (d) Sexual behavior. One of the patterns is sexual chasing, in which the male (black) dives at the quietly perching female (streaked, brownish) and she takes flight. Actual mating is usually initiated by the female. (e) and (f) Care-giving and care-soliciting behavior. The female inserts insects into the mouth of the nestlings, whose pattern of head raising and gaping is a signal for care and attention. (g) Allelomimetic behavior. As they fly in flocks, the blackbirds closely follow each other's movements, thus the flock moves as a coordinated whole. All the birds stay fairly close to the front with the result that the red-wing flock is spread out in a long rank perpendicular to the direction of flight. Investigative and eliminative behavior are not illustrated.

havioral systems are quite limited. Nor are these systems necessarily clearly distinct in every species. Behavior evolves in different ways in different animals, and the occurrence of both behavior patterns and their organization into behavioral systems must be determined empirically in each new species.

GENERAL SYSTEMS OF ADAPTIVE BEHAVIOR

Ingestive behavior. As a flock of blackbirds feeds in the fields, a first and very important kind of adaptation can be recognized. This might be called by the simple name of "eating behavior" except that in English we ordinarily make a distinction between eating solid food and drinking. Many animals, however, get most or all of their food in liquid form, including young mammals nursing at the breast, and young pigeons, which live on liquid "crop-milk." Since the same organs are involved in eating and drinking and the behavior is very similar, it is convenient to use a broad term, *ingestive behavior,* which includes both.

Shelter-seeking. Another general type of adaptation is seen as blackbirds come to roost at night. Small flocks fly into the trees near a cattail marsh, then down into the reeds, where they spend the night. At the peak season in August or September there may be 2,000 or 3,000 birds per acre on the roost. This tendency to seek out optimum environmental conditions and to avoid dangerous and injurious ones is found in almost all animals and may be called *shelter-seeking.* In its simplest form, which may be seen in one-celled animals under the microscope, a paramecium simply moves around until it finds the spot in the water drop which has those conditions most favorable to itself. As the drop dries up and conditions become increasingly unfavorable, the best place is sometimes close to the bodies of other paramecia, resulting in the formation of an animal aggregation. This is actually a very primitive type of social behavior, which can be called "contactual behavior." For most species of animals, the bodies of their fellows provide some form of protection against environmental changes, and it is possible that higher types of social behavior had their origin in this simple adaptation.

Agonistic behavior. One of the first things which the male blackbirds do upon their return from the south is to start fighting off intruders. The eventual result of the fight is that one of the

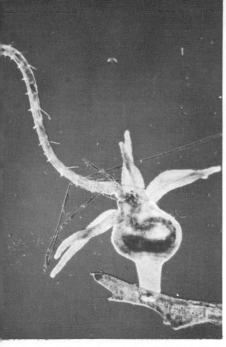

Ingestive behavior in a hydra (half-inch high, fresh-water coelenterate). The hydra swallows a long aquatic worm (with bristles, at upper left) through its circular mouth; the part ingested can be seen through the thin body wall of the hydra. (Photo by P. S. Tice)

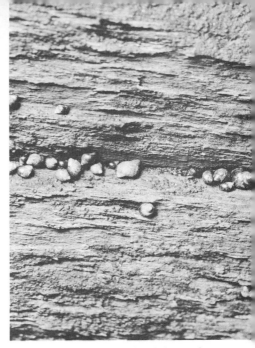

Shelter-seeking in rock crevices helps seashore animals to keep from drying out. *Littorina neritoides* (⅛ inch high) usually lives well above high-tide mark and is wetted only by spray. (Photographed at Trevone, England, by R. Buchsbaum)

Agonistic behavior in a herd of buffalo at Wind Cave National Park. Two bulls spar with each other at the edge of the herd. (Photo by J. A. King)

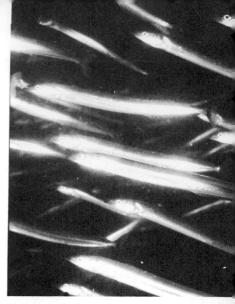

Investigative behavior is generally inconspicuous in animals with good vision or hearing. This mule deer is exploring its environment merely by standing still and directing its eyes and large ears in the direction of the disturbance. (Photo by J. W. Scott)

Allelomimetic behavior, in which members of a group react to and follow each other's movement, is strikingly shown by any school of fish. (Photographed in Roscoff, France, by R. Buchsbaum)

Care-soliciting behavior in young hermit thrushes. The young birds are unable to care for themselves, and this "gaping reaction" substitutes for adult food-gathering. (Photo by Hal Harrison)

Care-giving behavior of birds involves many different kinds of activity. Here a robin cleans the nest by removing a pellet of excrement deposited by one of the nestlings. (Photo by Hal Harrison)

birds escapes from the other. The behavior of both may be included under the general term *agonistic behavior*. In other animals the defeated individual does not always attempt to escape. When mice fight, one of the pair may simply hold up his paws in a defensive way or become completely passive. Any sort of adaptation which is connected with a contest or conflict between two animals of the same species, whether fighting, escaping, or "freezing," may be included under this term. The word "agonistic" comes from a Greek root which means "to struggle."

Sexual behavior. Once the blackbird males have established their territories by agonistic behavior, the females arrive and there begins the rapid pursuit which is characteristic of the courtship behavior of perching birds. Finally the female accepts the male and actual mating takes place. There is no difficulty in recognizing this behavior as *sexual behavior*, which includes courtship, coition, and any related behavior.

Care-giving behavior (epimeletic behavior). The natural consequence of sexual behavior is that young blackbirds eventually appear. They are at first unable to take care of themselves. This care is given by the parents, and begins long before the eggs hatch. After mating, a nest is built, the eggs are laid, and the female sits upon them until they hatch. Then the young are fed, chiefly on insects, which both parents catch. All this behavior can be described as the giving of care and attention, and by some authors it has been called "attentive behavior." It could also be called "maternal behavior," except that the male parent also takes part. The term "parental behavior" is a better one, but when we study other species of animals we find that care and attention are frequently given by animals that are not parents or, indeed, may never become parents. In the case of a colony of bees, the care of the young is accomplished by the workers, which are sterile females and incapable of being parents. Therefore, we can best use a broad general term, *epimeletic behavior*, which comes from a Greek word meaning "care-giving." Other more specialized terms which are sometimes used are "nurturant behavior," applied to rearing the young, and "succorant behavior," the giving of help to a distressed individual.

Care-soliciting behavior (et-epimeletic behavior). After the young blackbirds are hatched, they themselves begin to show some of the major patterns of behavior, and one of the first of these is ingestive behavior. Along with this they begin to do things which are not characteristic of adults—making cheeping noises and holding their heads up in the air with gaping beaks. This behavior consists essentially of calling for care and attention. Again we have the difficulty that there is no simple English word which describes this kind of behavior. It might be called "infantile behavior," except that in other species it is frequently found in animals which are completely adult. It seems to arise in situations in which an animal is incapable of adjusting or adapting by itself and substitutes for adaptation a call or signal which may result in care and attention from another animal. The word *et-epime-letic* is formed by prefixing a form of the Greek word *aeto* ("to call") on the word "epimeletic." This is a very appropriate name because the two kinds of behavior occur together frequently, and it means calling or signaling for care and attention.

Eliminative behavior. The young birds release the feces and urine in pellets surrounded by a membrane, and the parent birds lift these out of the nest as part of their care-giving behavior. There is no special behavior pattern associated with their own elimination, and this is also true of a great many other animals. However, certain young hawks have the habit of flipping the tail at the moment the feces are released so that the material is tossed out of the nest. The general effect is to keep the nest clean. Other animals also have special patterns of behavior connected with elimination. The tendency of cats to dig holes and bury their excrement is a well-known example.

Allelomimetic behavior (mimesis; contagious behavior). After the young birds leave the nest, and particularly later in the fall, they show a strong tendency to fly together in flocks with wonderfully integrated movement. The long axis of the flock is at right angles to the direction of movement, so that the flock itself is roughly shaped like a great bird. As each bird flies along, it does the same thing as those on either side. Both old and young show this behavior and spend a great deal of time doing it. Buffalo show the same sort of behavior as the herd moves from one grazing ground to another, and so do schools of fish. What sort of general term can we use to describe it? It is sometimes called "contagious behavior," but this has the unfortunate implication that the behavior is transferred from one animal to another like a disease. It is also called "gregarious behavior," but this is a

Fig. 3. Allelomimetic behavior in a flock of geese. The V-shape headed by a single individual is typical of these birds, which also do a great deal of vocal signaling or "honking" as they fly, often at great heights and during the night as well as the day. (From a photo by J. P. Scott)

more general term that implies only that animals come together and disregards the complex coordination of behavior in most flocks and herds. It could be called "imitative behavior," except that to most people imitation implies learning, and learning is not necessarily involved. It certainly cannot be strongly involved in the schooling behavior of young fish which are raised in hatcheries entirely away from adult fish and which still show a tendency to swim in coordinated groups. It could be called "mimetic behavior," except that this term has been used to describe situations in which there is a model whose activities or appearance is more or less independent of the mimic, as in the case of the hawk moth, whose behavior and appearance mimics that of a hummingbird. In the flocking behavior of blackbirds and in other similar behavior in herds of mammals and schools of fish, it is almost invariably evident that the animals are responding to each other rather than to a single model. Even when a leader is present, it always pays some attention to the behavior of the followers. The behavior always seems to involve some degree of mutual stimulation, and the term *allelomimetic* has been made up of two Greek words meaning "mutual" and "mimicking." It may be simply defined as behavior in which two or more animals *do the same thing*, with some degree of mutual stimulation and coordination.

Two kinds of behavior can easily be mistaken for allelomimetic behavior. A number of animals may happen to respond to the same stimulus at the same time, as when a group approaches a water hole and drinks. Close observation will usually determine whether or not the animals actually coordinate their movements. Then there is the case of a group of young animals in a litter, such as mice. They engage in the same type of activity and obviously stimulate each other, but observation will again show that their movements are not really coordinated.

Allelomimetic behavior may have many subsidiary adaptive functions: finding food, increasing the effectiveness of attacks by predators, such as wolves on a prey animal, and conversely, facilitating escape by the prey. Its most common function, however, is the mutual protection afforded by a group reaction to danger, providing safety. As a system, its general function is the facilitating of any response whose adaptiveness is increased by cooperation.

Investigative behavior. This kind of adjustment is not particularly obvious in blackbirds, which have very good eyes and can observe their surroundings at a glance. When, however, a mouse or a rat is placed in a strange box, he will go over it inch by inch with his nose and whiskers. Sometimes this is called "exploratory behavior." Animals which have hands, like raccoons or monkeys, will pick up a strange object, turn it over and over as they look at it and feel it, and they may also smell it or taste it. All this may be called *investigative behavior,* defined as any kind of sensory investigation of the environment. This system of behavior has a special interest for scientists. It is with this kind of adaptation that scientists respond to their environment, and they have developed highly elaborate and systematic ways of investigating the world around them.

As the above systems of behavioral adaptation are reviewed, it will be seen that all of them have some possibility of being involved in the reactions between two members of the same species. When this happens it is called *social behavior.* Ingestive behavior may be a purely solitary process, but many animals feed their young; and in the social insects mutual feeding is a common occurrence. Eliminative behavior may have social connotations when it is used in the marking of territories. Agonistic behavior, sexual behavior, care-giving, care-soliciting, and allelomimetic behavior are almost entirely social, involving reactions and adjustments to other individuals of the same kind. Finally, investigative behavior and shelter-seeking may or may not be social, depending on the circumstances.

By listing the systems of adaptation found in the behavior of the red-winged blackbird and comparing them with those found in other species, we have seen that most animal behavior may be described under nine principal headings. Not all of these are equally important. Ingestive behavior, shelter-seeking, and sexual behavior, being related to the fundamental life processes of metabolism and reproduction, are almost universal, as is investigative behavior, which is an essential part of behavioral adaptation. Care-giving—often accompanied by care-soliciting—agonistic, and allelomimetic behavior are found only in the higher animals, chiefly in arthropods and vertebrates, probably because these activities require either recognition of individuals through highly developed sense organs, or highly coordinated motor activities, or both. Eliminative behavior has minor or no importance in most species, except for land-living animals dwelling in nests or lairs, or where it has taken on a secondary signaling function.

Are these nine systems distinct from each other? As observations of the blackbird and other species show, these systems do occur independently in time. Allelomimetic flocking occurs in one season in the blackbird, sexual behavior in another. On the other hand, complex activities may involve a close succession of different systems; thus a parent bird may first show investigatory behavior as it searches for food and later epimeletic behavior as it feeds the young. Certain kinds of outright combinations between the systems are also possible, provided the behavior patterns do not interfere with each other, as in the case of group grazing in a flock of sheep, when ingestive and allelomimetic behavior go on at the same time.

Much of overt human behavior can be assigned to these nine general systems. We daily see ingestive, investigative, shelter-seeking, care-giving, care-soliciting, and allelomimetic behavior in those around us. Less commonly we see agonistic behavior in fights and quarrels. Sexual and eliminative behavior are partially concealed according to social custom. We only get into difficulties when we look at the highly symbolic verbal and creative activities that are a specialty of the human race, partly because there is often very little overt activity involved, and partly because verbal activity, like any communication, can be a part of any behavioral system. Incidentally, no scientist has yet done a serious observational study of the behavior patterns and behavioral systems of the human species.

The list provides a convenient guide for the description of behavior in a new species, but in any particular case certain behavioral systems may be absent. Adult sponges show little or no behavior of any kind. Ingestive behavior is absent in parasites such as the tapeworms, and certain species of insects completely lack sexual reproduction and hence sexual behavior. On the other hand, there is always the possibility of discovering an entirely new system.

The kinds of behavior found in any animal depend a great deal upon its environment. An animal which is given every care in a small laboratory cage has very few problems to meet in ordinary living and consequently will show only a few kinds of adaptation. The best way to make a primary study of animal behavior is therefore under natural conditions.

OBSERVATION AND DESCRIPTION OF BEHAVIOR

Some fortunate scientists have had the opportunity of observing animals under completely natural conditions: Fraser Darling, who followed the red deer of the Scottish moors, C. R. Carpenter, who studied the behavior of howling monkeys on Barro Colorado Island in the Canal Zone, and T. C. Schneirla, who followed the wanderings of army ants in the same locality. Such opportunities have recently become more common, but they are rare in any scientist's life. Luckily there are animals closer at hand which can be effectively studied.

A number of years ago we lived on a small farm in Indiana and bought a couple of sheep in order to keep down the weeds and grass in the pasture. There was more than enough room for the sheep, and they could be allowed to wander around, feeding and taking shelter at will, with very little human care or interference. As lambs appeared and the flock grew, their behavior became more and more interesting and enabled us to make a systematic study which illustrates various methods for observing natural behavior.

Methods of observation. The best way to observe animals is to get into a position where they are not disturbed by your own behavior. Sometimes this can be done from a blind, but it is even better if the animals become accustomed to you and no longer react fearfully to your presence. With great patience this can be done even with a wild species, as Jane Goodall did with chimpanzees, and Phyllis Jay with langur monkeys. It was very easy to do with the flock of sheep that lived near us, as they soon became indifferent to anything that we did. We then began taking notes on their behavior.

We found that one of the best kinds of observation is a short ten-minute sample of behavior. This is short enough so the observer can give full attention to the job; and if the animals are doing anything particularly interesting he can watch longer. If they happen to be inactive, little time is wasted. The observer tries to write down everything that the animals do. It is also a good idea to keep a record of general weather conditions, such as the wind and temperature and whether it is cloudy, sunny, or stormy. And of course one should always note down the place, the date, and the time. Some workers prefer to use a portable tape recorder for making notes. This has the advantage that one does not need to look away from the animals and there is less chance of missing something. On the other hand, there is less chance of catching a mistake in recording, and the data must eventually be transcribed. Another disadvantage is that the sound of the observer's voice may disturb the animals unless he uses a sound-absorbent mask.

As one watches, various ideas and possible explanations of the behavior will come to mind. These should be noted down too; but since they are not facts, but inferences or conclusions, they should be marked in some way, such as inclosing them in brackets. The following is a sample observation of the behavior of the sheep. At this time the flock consisted of several older sheep and one new lamb, born a few days before.

5–19–41. 6:30 P.M. New male lamb born to young black ewe. Staggered around, trying to suck on mother, which occasionally kicked him out of the way. An older lamb [12 days old] away from own mother with two other ewes. Showed much interest in newborn lamb, attempted to mount its mother, was butted away. Also attempted to butt with newborn, which did not seem disposed to do so, but apparently attempted to suck older lamb. Latter also attempted to mount own mother from front. [Very precocious; smell seemed to act as stimulus to sex behavior.]

The motion picture is another technique which has considerable use in description. One can never reduce behavior to words in a completely satisfactory fashion. After first obtaining some idea of what the animals are doing, it is often helpful to make moving pictures as a supplementary record. A 16-mm. camera is the best type, and a telephoto lens is essential. Cameras are designed for close-up views, and the details of behavior are lost in distant shots. Also, if one is too close to the animal the noise of the camera will disturb its behavior. Even without a camera it is usually a good idea to observe at a distance. A pair of good binoculars is essential for work with most animals, and in some cases a telescope is useful. Television recording on tape has some advantages, as both sound and sight can be recorded and the tape re-used, but this is rarely feasible under field conditions. In any case, the resulting picture is never as clear as those that can be obtained with photography.

The daily round. As we watched the sheep it became obvious that they did different things at different times of the day. If we made our observations at the same hour each day, we saw only a few types of behavior often repeated. It was therefore necessary to make at least one full day's observation. Other animals which are active at night may require a 24-hour observation period. From this one can learn the times when the animals are most active and concentrate observation in those periods.

In the case of the sheep we found that in the colder weather they spent the night bedded down in a barn or shed but would come out and begin to be active soon after it became light. They moved around the field for two or three hours grazing, then stopped to lie down and chew the cud. This cycle was repeated throughout the day until the sheep finally bedded down for the night just before sunset. They usually followed a fairly regular path covering most of the field. At different times of year their behavior would vary somewhat. Bad weather would inhibit their activity. In the summertime, when many biting insects were out, they often stayed in a dark cool place during most of the day and did their feeding and wandering at night.

From this study we saw that there were two predominant systems of behavior in the sheep: *ingestive behavior* and *allelomimetic behavior*. The flock spent almost all its time either grazing or lying down to chew the cud, always in coordination with other members of the flock. In either very hot or very cold weather *shelter-seeking* was also evident.

Circadian rhythms. In other animals activity may be more precisely related to time. As biologists long ago discovered, rodents placed in an activity wheel become active during certain hours, usually at some time during the night; the clever investigator of rodent behavior in the laboratory reverses the day-night cycle so that he can watch animals at an hour convenient for himself. However, if the rodents are placed in complete darkness as far as possible, isolated from all outside stimulation, they still continue to show rhythms of behavior at roughly 24-hour intervals. These have been called *circadian rhythms* (literally meaning "about a day"), and they occur in many different kinds of animals, invertebrates as well as vertebrates. One of their peculiarities is that under conditions of total darkness or constant light, the biological clocks of some animals run fast and those of others run slow. Exposure to a normal day-night cycle results in the biological clocks being reset.

The seasonal cycle. It was obvious to us that the behavior of the sheep varied from season to season, and that we could get a complete picture only by following them through a whole year. As we did this, we saw other types of behavior. During the autumn the amount of *sexual behavior* greatly increased. The males became much more interested in the females and the latter came into heat during this season. This period of estrus or sexual receptivity lasted only one or two days and was not repeated after the females became pregnant. Since the females will not react to the males except during estrus, their sexual behavior was extremely limited. During the mating season there was some fighting between males who attempted to compete for females in estrus. This was the only time when much *agonistic behavior* was evident.

The gestation period of the sheep is approximately five months, which meant that in the spring the lambs were born and the mothers began to take care of them. The mothers licked and nosed the newborn lambs and allowed them to nurse. If small dogs came near, the mothers would try to drive them off. This *epi-*

meletic or *care-giving behavior* occurred only in the females, and the rams paid no attention to the offspring. At the same time there was an increase in *care-soliciting behavior*, as the lambs called to their mothers when cold, hungry, or alone.

Regular seasonal cycles of behavior are common among the higher animals. Sometimes these are adaptations to changes in climate, and they are also frequently associated with reproduction. The lower invertebrates may show even more extreme seasonal changes. Spectacular examples of seasonal sexual behavior are found in certain marine worms. The palolo worm of the South Pacific, for example, spawns only on special days in October and November as the moon reaches its last quarter. Similar spawning habits are seen in the nereid worms of our own Atlantic Coast, but at different times of the year. Seasonal migrations and hibernation also produce yearly cycles of behavior more extreme than anything seen in the sheep.

As with daily cycles of behavior, there are also seasonal cycles governed by internal biological clocks. The woodchuck, or groundhog, a common rodent of meadow lands in the eastern part of the United States, hibernates, staying in holes all winter long and emerging in the late winter. David Davis kept woodchucks under laboratory conditions where there were no cues to outside changes in light and temperature and found that it was almost impossible to disturb the cycle. Under natural conditions in Pennsylvania male woodchucks emerge so regularly on or near February 2 that this has become known as Groundhog Day.

Development of behavior. The young lambs did not behave exactly like the adults, and in order to get the complete picture of behavior we had to follow their activity from birth until they reached maturity. One of the first activities was their special pattern of ingestive behavior, sucking from the mother's udder while thrusting against it and wriggling the tail. After a while it became obvious that the mothers always allowed the lambs to nurse just after they came when called. This appeared to train them to follow their mothers.

Within a few days the lambs began to eat grass but spent much less time grazing than did their elders. Instead they spent a good deal of time in play activities. As we watched, we saw that most of the play consisted of immature forms of adult behavior patterns, but directed toward no particular end. The young lambs would run and leap together in a playful form of allelomimetic behavior. Occasional mounting was obviously a playful form of sexual behavior. As they grew older their behavior became more and more like that of adult sheep.

In order to study the development of behavior more closely, two young lambs were taken from their mothers at birth and fed from a bottle for several weeks. This is an excellent technique to use with young animals because it is often very difficult to study them closely in any other fashion. In the case of the bottle lambs, this experience altered their behavior in that they seldom showed any tendency to follow the flock. They developed into very unsheeplike sheep—fearless, and independent of their own kind. This indicated that a great deal of what we think of as sheeplike behavior is not necessarily inborn but is in part a product of the social environment.

This kind of systematic study of adaptation needs to be done for a large variety of animal species in order to lay a firm foundation for general conclusions regarding behavior. Relatively few species have ever been studied in more than a superficial manner, and the information is particularly important at the present time because of current interest and progress in the study of social behavior and organization. Each species which is studied in this fashion provides the basic data for understanding the workings of animal societies. Any intelligent observer with good elementary training in biology and psychology can make a contribution along these lines. It is to be hoped that the description of animal societies and the kinds of behavior on which they rest will soon be as popular with biologists as the description of new species once was, and indeed great progress is now being made with field studies of primates.

THE EXPERIMENTAL ANALYSIS OF BEHAVIOR

As we observe animals in the field, certain ideas or theories occur to us which seem to explain their behavior. These ideas can occur so frequently that we have to be careful not to substitute them for our observations. We should always remember that a theory rests on a firm foundation only after it has been tested by experiment.

Field experiments. In working with real situations, the best ex-

Fig. 4. A field experiment with cliff swallows. *On the left,* two birds are shown brooding their eggs in adjoining nests, as they would look if we could see through the outer mud wall. The birds are completely peaceful. *Right,* during the absence of the two birds, the experimenter has broken down the

periments are done under field conditions with only a slight degree of interference in the lives of the animals. J. T. Emlen did such an experiment with the behavior of cliff swallows. These birds build their nests out of mud in protected spots under cliffs and the eaves of houses. Unlike blackbirds, which drive others away from a large area around the nest, the cliff swallows build their mud homes side by side in large colonies. Emlen wanted to find out whether these birds showed any tendency to defend their own nests, but in order to do this he had to make sure that the same birds always came back to the same nests. All the swallows looked very much alike, so he sneaked up on them with a water pistol loaded with some quick-drying paint and squirted a few drops on their feathers. They became speckled in various patterns which were easy to tell apart, and he found that the same birds always did come back to the same nests. Very little fighting could be observed between the owners of neighboring nests. Then he began the real experiment, which was to tear down part of the wall between two adjoining nests. Whenever the birds met in the combined nest, they started to fight. Within a day or two they repaired the hole and lived peaceably as before. Emlen came to the conclusion that the swallows had very small territories around their nests and that these were usually well respected by the other birds, with the result that little fighting occurred.

This simple experiment illustrates one extremely important technique in field experiments which arises out of the nature of experimental design. An experiment always assumes that there is a cause and effect relationship, and that changing the cause will change the effect. This must be repeated a sufficient number of times so that one is sure that the results do not occur by chance.

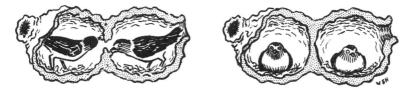

wall between the nests, and the birds begin to fight each other. *Left,* the birds begin to repair the wall. *Right,* the birds are once more peaceful. (Sketches of an experiment described by J. T. Emlen)

Repeating an experiment under natural conditions is always a difficult business, and one of the absolute essentials is that the animals being experimented upon are marked or identified in some way. Otherwise one may repeat the experiment on different individuals with unexplainable results.

Laboratory experiments. Field experiments have the advantage of being realistic, but there are always all sorts of interruptions and variable environmental factors which interfere. It may rain one day and be clear the next, or the experimental animal on which many hours have been spent may suddenly disappear as the victim of a predator. All these things can be better controlled in the laboratory, and laboratory experiments always have the advantage of greater possible accuracy.

One of the basic theories that arises out of observation and experiments on behavior is that behavior is affected by multiple factors. This is particularly evident in experiments with heredity and environment. In our laboratory we noticed that young puppies of different breeds seemed to respond differently to human beings, some being friendly at the start and others being fearful and aggressive. We therefore raised litters from several different breeds of dogs, giving them the same kind of room in which to live and the same kind of care and nutrition. The next step was to devise a test of their reactions to people. The simplest thing was to have the experimenter do the same sorts of things which people ordinarily do to puppies: walking toward them, patting them, stooping down to them, and calling them. After some practice, this could be done in a standard way; and the reaction of each puppy was checked off after each act of the handler.

The experiment now had to be gone over for stray environmental factors. One possibility was that the puppies might respond differently to different experimenters, and the test was therefore arranged so that two different people tested the same puppies. Another was the order of testing. As a litter was being handled, the first puppies saw action much sooner than those which were kept waiting around to be tested last, and this might affect their behavior differently. Therefore the test had to be repeated, reversing the order. However, if the test was repeated too often, the puppies would be likely to form habits as a result, so the test was repeated only once every two weeks, with the expectation that the puppies would have largely forgotten their earlier experience.

Other miscellaneous things which might have affected the puppies were the time of day and the temperature. An attempt was made to repeat the test at the same time of day and to maintain a relatively uniform temperature in the room.

As the test went on, it was apparent that not all the puppies were getting the same amount of milk from their mothers and that puppies in smaller litters were obviously better fed. The hungrier puppies seemed to pay more attention to people, and the only legitimate method was to compare behavior of litters of the same size.

When all the possible environmental effects had been considered, we concluded that there were definite differences in behavior which were due to heredity. In addition, repeating the experiments showed that there were great changes in behavior in early life. The puppies started out by being relatively fearful and quickly became more friendly. The experiment not only led to the conclusion that there were hereditary differences but also suggested two theories for further experiments: that the state of nutrition affected the social behavior of the puppies, and that there was a particular time in life when the puppies could be very easily taught to be friendly.

In designing an experiment on behavior, we must always remember that there are many factors which can alter behavior and that these are frequently unknown at the start of the experiment. To avoid misleading results, the following precautions should always be taken:

1. *Time*. The time of day and time of year are always impor-

tant, and it is best to plan an experiment so that experimental and control animals are tested almost simultaneously.

2. *Order.* Order is related to time. If the experiments cannot be done simultaneously, it is best to reverse the order in experimental and control animals or to rotate the order in some way on repeated tests.

3. *Individual differences.* Every individual behaves differently from every other one, partly because of heredity and partly because of differences in previous experience. The ideal experiment is one in which the same individual is both experimental and control, but this is not possible if it is necessary to observe experimentals and controls simultaneously. One of the favorite types of experiments is therefore that of paired comparisons, in which an experimental and control animal are matched as accurately as possible in a pair. Heredity can be controlled by using inbred strains and the same sex. Previous environmental factors can be fairly well matched by taking animals which are the same age and born of the same parents. This is easy to do with animals which produce more than one offspring at the same birth. However, it is not necessarily true that two individuals in the same litter have had identical environments, as the social environment depends upon interaction between the animals making up the group, all of whom show individual differences.

4. *Social environment.* This can be kept constant by raising animals in isolation except when this itself changes behavior into something quite different from that of the normal run of individuals. It should also be remembered that animals frequently react toward an experimenter as part of the social environment and that their relationship with him may affect the outcome of the experiment. Even laboratory rats will perform better on mazes if they are accustomed to handling ahead of time rather than being suddenly seized by a total stranger.

5. *Physical environment.* Factors such as temperature, light, and noise are fairly obvious and easily controlled. Other factors may be less evident, such as the physical position of the animals, air currents, changes in humidity and barometric pressure, and even odors and gases in the atmosphere. In one study of the effects of drugs on the behavior of mice, the experimental animals were placed in the same box as the controls with only a wire parti-

tion between. All physical factors seemed to be equal until it was noticed that the light came from a window in one side of the room and that the mice on the lighter side of the box ate more than those on the darker side.

6. *Accuracy of the experimenter.* Anyone is likely to make mistakes, and even the most experienced scientist is not exempt from this failing. The sorts of things which are measured in behavior are particularly liable to error. When the chemist weighs something he can go back and look at the scale several times to check his observation. However, when we record the duration time for a particular behavioral reaction, it is not possible to go back and look at it again after the behavior has stopped. If a mistake in timing has been made, it will remain on the stop watch. Likewise, if an observer happens to look away at the time an animal does something, he may quite honestly maintain that the animal never did it. Various ways have been worked out to check the accuracy of an experimenter, and one of the best of these is to have the recording done independently by two different individuals and to compare the results. These will never be completely identical but should give a high correlation.

It will be seen that any carefully done experiment on behavior becomes a very complicated affair and that it usually is not possible to get conclusive results with a single simple experiment. The experimenter has to do many careful repetitions of his work, varying each possible causal factor, before he can be sure that he has a real understanding of behavior.

Careful experimental technique succeeds best when it is based on a thorough knowledge of the fundamental behavioral adaptions of a species. Most animals adapt to an unpleasant situation by attempting to escape, but not always in the same way. In our laboratory an experimenter attempted to test the learning ability of mice by putting them in a box with an electric grid for a floor. A buzzer was always started before the experimenter turned on the electric current, so that the mice could escape by jumping to the other side of the box where the current was turned off. The experimenter reported that, unlike rats, the mice simply would not learn, because he could repeat the experience two or three hundred times, and the mice still would not escape. Then another observer happened to notice that the mice always tried to climb out of the cage instead of running to the other side in order to escape. He arranged the box so that the mice could climb the sides, and with this new arrangement the mice learned to climb out of the way at the sound of the buzzer in only one trial.

It looks as if mice and rats, which appear so much alike on the outside, have fundamentally different ways of getting out of a painful situation. Such differences between species must have a physical basis. Each animal has certain behavioral capacities, the most obvious of which are based on body structure, and this is the subject of the next chapter.

SUMMARY

The science of animal behavior is primarily concerned with *what an animal does*. Each kind of animal has characteristic ways of adapting to change, known as behavior patterns. Such a segment of behavior having a specific adaptive function forms a natural unit of behavior, in the sense that natural selection can only act where there is adaptation. Behavior patterns in turn are organized into behavioral systems. As many as nine independent systems can be found in a species, and these provide a convenient guide for systematic observation. Each species not only has a characteristic combination of behavior patterns but shows its own peculiar emphasis on certain behavioral systems. The red-winged blackbird shows all the major systems, but other species may not show such a variety of reactions.

The best observational studies are done under natural or semi-natural conditions, using short-sample techniques. These may be used to study the daily round, seasonal cycle, and individual development of behavior. The results of such a study on the behavior of sheep are described. We need systematic studies of many representative species in order to lay a firm foundation for the science of comparative animal behavior. Each study of this sort raises exciting new questions which can be answered by experiment.

There are two general methods for studying behavior: observation and experiment. Neither provides all the answers by itself. Ideally, the observer and the experimenter form a partnership. The observer accumulates basic facts which suggest theories and explanations which can be tested by the experimenter. Their combined efforts may result in more refined theories and more accurate facts. In this chapter we have emphasized the need for systematic and thorough investigations of both kinds.

The basis for sound experimental work is a knowledge of the behavioral capacities of an animal. What an animal does in any type of adaptation depends upon how it can use the sort of body with which it is born. The wide variety of behavioral capacities in the animal kingdom is the subject of the next chapter.

CHAPTER 3

Differential Capacities
Anatomy and Behavior

Guinea pigs are in many ways singularly uninteresting animals. They have only a few simple patterns of behavior for responding to any sort of stimulus. The high point of the day for a laboratory animal is feeding time, and when we give guinea pigs fresh hay they run through it squeaking with excitement, but a few minutes later they are sitting on top of it, stolidly munching away. When reared outdoors they build no nests or burrows. Under any conditions they take very little care of their young except to allow them to nurse. They cannot fight effectively against anything except another guinea pig because their necks and backs are not flexible enough to permit free biting and scratching. This makes them docile pets. When you pick up a guinea pig, it scarcely bothers to struggle and hangs limply in your hand. Even their sexual behavior is simple, with no elaborate patterns of courtship. Finally, comparative psychologists find that it is difficult to set up tests on which a guinea pig will show any sign of learning or intelligence. We begin to wonder how the wild ancestors of guinea pigs could have survived at all.

One possible explanation of their oversimple behavior might lie in their sense organs. Being nocturnal animals, they might have poorly developed eyes and hence be unaware of much that goes on around them in the daylight. When we look into this, however, we find that they have moderately good eyes and very good ears with which they can respond to high-frequency noises beyond the capacity of the human ear.

A better explanation of their simple behavior seems to lie in their limited motor apparatus. The legs of a guinea pig are adapted for running and nothing else. Consequently, there is little more that a guinea pig can do. It does most of its living at night,

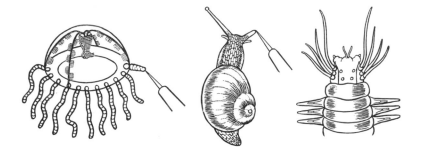

Fig. 5. Tentacles and other tactile organs are found in a variety of animals. *Left to right,* jellyfish, snail, head of nereid worm. *Opposite page,* catfish and kangaroo rat. (Original)

and its wild ancestors undoubtedly escaped from predators chiefly by running away in the dark. Other than this the guinea pig spends its life eating and reproducing its kind.

This example illustrates the point that one important factor affecting behavior is the kind of physiological and anatomical capacities with which a species is endowed. Since behavior is an attempt to adapt to changes in the environment, the kind of sense organs which an animal has determines the changes to which it can respond, and its motor organs determine the kinds of responses it can make.

SENSORY CAPACITIES

Animals react to a great variety of stimuli, which come from far and near. The special organs which receive these stimuli may be conveniently grouped according to the distances at which they are effective. The tactile senses receive stimuli from the part of the environment which directly touches the animal. The chemical senses of odor and taste extend the range of response, but this is still limited by slow rates of diffusion and interference by water and air currents. Finally, sight and sound enable an animal to respond to objects far beyond the microclimate in which it lives. In the case of sight, responses may be made to the moon and stars, which are almost limitless distances away.

Tactile senses. Amebas have no special sensory structures and yet withdraw from touch and heat. The free-swimming paramecia

add a response to gravity but have no specially developed tactile organs. It is only in the coelenterates that we first find a cell with a special part used for touch. Jellyfish and hydras catch their prey and sting them with their tentacles. Each stinging cell has a small hair which responds to either touch or chemical stimulation, causing the discharge of a poisoned barb into the prey. The action of the hair increases sensitivity. A slight push easily moves the tip, which acts like a lever and transmits even a very small stimulus to the cell. Similar sensory hairs are found in many higher animals. When we look at an insect like a fly under a microscope, we see that it is covered with sensory bristles. In some animals sensory hairs are long enough to make contacts at a distance from the body. In nocturnal mammals like mice and cats the vibrissae, or nose hairs, are used to feel out objects in the dark.

The coelenterates also show the first true tactile organ, the tentacle. Every polyp and jellyfish has a circle of these around its mouth. When we touch one with a needle, it quickly contracts, showing sensitivity to contact. The fleshy tentacles on the head of a marine nereid worm react in the same way. The "horns" of snails are really tentacles which also carry the eyes. The flexible antennas of arthropods are another kind of tentacle and are frequently developed to great size in animals which are active in the dark. In insects like katydids they are as long as the whole body. Even in vertebrates there are a few animals, like catfish, with tentacle-like organs. All of these organs are "feelers" and are particularly important to animals which either live in dark places or have no eyes.

Reactions to gravity are closely related to the tactile sense. Some jellyfishes have at the base of the tentacles a small organ

called the lithocyst. It contains a freely moving granule which falls
on the sensory hairs around it and thus presumably tells the animal which side is up. It is interesting that all sense organs which
react to gravity apparently use such sensory hairs. Crustaceans
like the crayfish have a small pit which they fill with sand grains
after each molt. If we keep a crayfish away from sand and give it
iron filings instead, we can do an interesting experiment. We put
a strong magnet near the animal as it lies in the water, and it turns
on its side or upside down, depending on the position of the
magnet. Even in the semicircular canals of vertebrates, the flow
of fluid which tells the animal of a change in position is picked
up by little sensory hairs.

All animals react to injurious tactile sensations, and higher
animals may have special nerve endings responding to the stimuli
of pain, heat, and cold. These, like other tactile senses, depend
on direct contact with the environment. Even the longest tentacles
reach out only a short distance from the body. The organs of
touch bring in a limited but very important group of stimuli, those
which come from the part of the world closest to the animal.

The chemical senses. Protozoans react to various natural and
unnatural chemicals in the water around them, but they have no
special organelles of taste. True organs of chemical sense are
common in multicellular animals. As a planarian travels along, it
delicately waves the projecting flaps on either side of its head,
tasting the water as it goes with its sensory lobes. In most invertebrates the tentacles or antennas serve as organs of chemical sense
or taste as well as touch. In land animals such as insects the same
organs are used for the detection of odor.

All vertebrates have nostrils. Fish use blind pits to taste the
water as they swim through it or force a current over their gills.
In amphibia the nostrils become connected with the mouth and
are used for breathing air. In these and other land vertebrates,
the nostrils are used for sampling chemical changes in the air
instead of in water. The power to detect the changes in liquids
has not been lost but is limited to the mouth and tongue. In contrast, fish have taste buds extended to the outside of the head.
Some fish have tasting organs even on the tail and can literally
taste with the whole body.

The possession of well-developed chemical senses considerably
extends the possible range of contact with the environment. In

water-living animals the distance is limited by the slow rate of diffusion of chemical substances. It is only when there is a current of water that the animal can receive stimuli from farther than a few feet away. The range for air-living animals is considerably increased by the more rapid diffusion of gases and the fact that air currents are much faster and more variable than water currents. Under favorable conditions moths can locate their mates by odor from a distance of half a mile. Nevertheless, chemical stimuli travel much more slowly than those of sound or light, which permit immediate contacts with the environment at greater distances.

Sensitivity to X radiation and electric fields. Human patients are normally unaware of X rays while undergoing diagnosis or treatment, but experiments with avoidance conditioning in rats demonstrate that rats can detect X radiation in very low doses in the range of 0.5 roentgens per second or less. One of the most sensitive areas is that of the olfactory bulbs in the brain, indicating that the X rays are directly stimulating the sensory fibers which normally respond to gases. However, rats can also detect X rays reaching other parts of the head and in the abdominal region, so other parts of the nervous system must also be sensitive. There is no indication that this ability is useful to the rat under natural conditions, and it is simply another case of capacities that were evolved for particular functions having additional uses as well.

Ever since the days of Galvani, experimenters have been stimulating nerves through the use of electrical currents. Such stimulation sets up a nerve impulse which is physiologically no different from that produced by natural stimulation. Animals also respond to external electrical stimulation, which may not only activate the sense organs directly but if strong enough may burn the skin or stimulate muscles to contract. Such capacities are not often functional under natural conditions except in the case of water-living animals. Certain fish can detect the presence of weak electrical fields produced by the muscular contractions of other fish, and some even produce such fields by their own muscular contractions and use these to detect distortions produced by objects otherwise invisible to them. Such sensory capacities are, of course, useful only at short ranges, but at least one freshwater electric fish, the Gymnotus of South America, has evolved a system of electrical signals that are employed during agonistic encounters.

Sound and light. All water-living animals are presumably sensitive to vibration in the water through their tactile sense organs, but very few of them have any special apparatus for this. Among land-living animals it is only the vertebrates which have highly developed sound-perceiving organs. Crickets and grasshoppers, which "sing" as part of their mating behavior, have a round plate on the side which functions like an eardrum, but these are the exception rather than the rule among insects. Most members of this class have simple sense organs which are stretched like fiddle strings from the outer skeleton, and these are sensitive to either sound or touch.

The ear of higher vertebrates was apparently developed first by fishes as a balancing organ. Sharks have large semicircular canals, and sound waves pass directly through the head to the lower part of the inner ear. There is nothing corresponding to the human external and middle ear. The tiny bones that in the human ear transmit vibrations from the eardrum to the inner ear are developed from structures that in the shark form parts of the jaw. Fish also have an external set of organs connected with the auditory apparatus, the lateral lines, which extend down the entire length of the body and send branches over the surface of the head. These organs are sensitive to movements of water, and with them the fish can detect moving or still objects (the latter, as water set in motion by the moving fish is reflected back from the motionless object) as well as the water currents in a stream. In addition, certain special cells in the lateral line organs are sensitive to weak electric currents and are presumably the principal sense organs involved in the detection of electric currents described above. The whole apparatus is so different from the external ear of land-living vertebrates that few people suspected that sounds might be important to fish until the invention of underwater listening devices for submarines revealed that the ocean is quite a noisy place when fish are around. They grunt and groan, and most seem to need no special organs to receive these sounds because water is a very efficient conductor of vibration. Anyone can test this latter point for himself by putting his head under water in a bathtub. The fresh-water carps and catfish have especially acute hearing. The air bladder is connected to the ear by a chain of small bones. This apparatus is superficially similar to the human sound-receiving organs but is directed toward the inside instead of the outside of the body. The land-living vertebrates have developed the eardrum, an external membrane sensitive to the weaker vibrations carried by the air. Many species of mammals also have external funnels which concentrate sound waves on the eardrum, and bats have unusually large external ears for their size. Their ears actually take the place of sight in a process of locating objects by echoes, and this will be discussed in detail in a later chapter.

Special organs for light reception are much more universally developed. The chlorophyll-bearing euglena has a light-sensitive spot which guides the animal toward optimum light conditions. Some flatworms and coelenterates also have eye spots or light-sensitive regions. These organs can only respond to light and dark and have no power of resolving the shape of objects. Image-producing eyes have been developed in only three phyla, the mollusks, arthropods, and chordates, and have taken a different form in each. Strangely enough, certain mollusks such as the squids have large eyes which are built on a plan similar to that of vertebrates and seem to function with almost equal efficiency.

Like vertebrates, the arthropods have lenses for concentrating light on the light-sensitive cells of the eye. However, there is a small lens for each nerve ending, and the eye becomes efficient only when hundreds of units can be combined together as they are in the compound eyes of many insects. Some of these individual ocelli have a short focus and some a long one, so that the animal can perceive whether objects are near or far. There is no power of focusing the lens as in the vertebrate eye, but the compound eye has other advantages. The individual ocelli can point in any direction, and the stalked eye of a crab can cover a field of vision of three-quarters of a full circle. Using two large compound eyes, dragonflies move at very high speeds, catch other insects on the wing, and avoid capture with great success.

The basic plan of vertebrate vision is two large single eyes with focusing lenses. There are traces of a small third eye in the top of the head, which in adults is well developed only in a rare New Zealand reptile, the tuatara. It is possible that this eye was used not so much for sight as for light-perception which controls the endocrine organs regulating the seasonal cycle of behavior; but

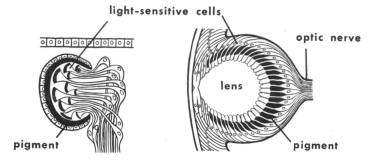

Fig. 6. Section through eye of a planaria, *left,* shows how pigment screens light-sensitive cells so that light is received only from one side. This enables animal to tell direction from which light comes. The nerve cells run to the brain. (From Buchsbaum after Hesse)

Section through eye of a nereid worm, *right,* shows advance over planarian eye. A gelatinous lens concentrates light upon the specialized rodlike ends of the pigmented cells that comprise the primitive retina. (From Buchsbaum, based partly on Kukenthal)

we have no evidence on this point. Vertebrate eyes are adapted for many special purposes, but their most essential function is their reception of stimuli at great distances. This is of particular importance to the larger species.

Since sight is well developed in only three groups of animals—the mollusks, arthropods, and vertebrates—we might conclude that highly developed sense organs are correlated with complex organization. However, in any particular species the capacities of the various sense organs may be very unequal, and it is unusual for all of them to be highly developed. A bird like the turkey buzzard has both excellent eyesight and a keen sense of smell, but in many birds the sense of smell is only slightly developed. As might be expected in animals whose mouth and tongue are inclosed in horny material, birds make little use of the sense of taste. In some nocturnal birds such as owls, the eye is highly developed for night vision, whereas some night-living mammals such as mice show a reduction of vision and a dependence on chemical and tactile sensations. The sensory adaptation of any species can be understood only after a special study of the sense organs, environment, and behavioral habits of that particular animal.

The senses of sound and sight permit contacts with the envi-

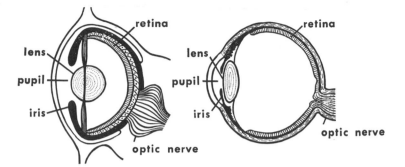

Section through eye of squid, *left,* shows structural similarity to human
eye. Lens forms an image on the retina. (After Buchsbaum)
Section through human eye, *right.* (After Carlson and Johnson)

ronment at much greater distances than do the chemical and
tactile senses. Animals which possess them therefore have the
possibility of developing much more complex types of behavioral
adaptation, provided they also have the necessary motor apparatus and coordinating system.

Other senses. There may be other ways, as yet undiscovered,
in which animals maintain contact with the physical world. One
possible source of stimulation is the earth's magnetic lines of
force, which change as an animal moves through space or changes
direction. Such a sense of magnetism would help account for the
so-called "sense of direction," and would explain the orientation
which birds are able to maintain during their long migrations.
However, no sense organ has yet been discovered which would
be capable of receiving this information. The pecten, a peculiar
fan-like structure in the eye of birds, has been suggested as a
possibility, but its location makes it very difficult to modify by
experimental procedures and there is still no evidence of its
physiological activity.

MOTOR CAPACITIES—LOCOMOTION

The systematic study of the motor capacities of an animal
begins with the observation of movement, starting with movements of each part of the body and proceeding to movement of
the whole body, or locomotion. In some cases movements are so
rapid that they cannot be observed with the naked eye and must

be measured with the help of high-speed photography. The wing movements of the humming bird are so rapid that they present only a blur, and even the wing movements of larger birds are difficult to follow. The movements of which an animal is capable do not have special functions and hence form only parts of complex adaptive behavior patterns. However, such capacities are one determinant of behavior patterns. The movements themselves are largely determined by the kinds of motor organs which the animal possesses and hence are closely related to anatomy.

Motor capacities are more directly related to adaptation through locomotion, since one of the most basic forms of adaptation is moving from an unfavorable environment into a more favorable one. There are large groups of animals which have little or no power of movement, particularly the sponges and many sorts of parasites. In these groups behavioral adaptation is unimportant or almost entirely absent, and most adjustments are made on a physiological basis. Among animals which do move there is a very wide variety of motor organs, each used in a special kind of adaptation.

Movement by pseudopods, flagella, and cilia.—These means of locomotion are the specialties of one-celled animals, the protozoans. Large animals cannot use such methods to move their bodies, but cellular locomotion may survive inside them. In the human body some white corpuscles may move with pseudopods in ameboid fashion, and the sperm cells of the male are propelled by flagella. Cilia are found in the linings of the human respiratory tract.

Cilia are short, flexible, hair-shaped extensions of the living cell. They usually occur in very large numbers, and in a protozoan they row the animal about by beating the water more strongly in one direction than another. The commonest inhabitants of any culture of swamp water are protozoans of the class Ciliata, ranging from the large Spirostomum, gliding about like ships, to the small hypotrichs which have their cilia fused into spikes on which they crawl among the algae like little fleas. Some of the many-celled animals, or metazoa, also move by cilia, and of these the walnut-sized ctenophores are perhaps the largest. These marine animals have rows of combs which are really fused cilia that beat in unison and produce an almost imperceptible movement as the animals float in the water.

Tiny flatworms of the class Turbellaria use cilia to glide smoothly and slowly on the bottom of an aquarium or even on the underside of the surface film of water. They are also able to crawl by muscular movements when necessary. The nemertean worms of the ocean may also move by cilia, and so do rotifers and a variety of other small water-living creatures. Ciliary locomotion is, then, limited to small soft-bodied animals which live in the water. Since cilia are projections of single cells and limited in size, their effectiveness is inversely proportional to the size of the animal.

Jet propulsion. This is a very old form of locomotion in the animal kingdom. In a pool back of the Marine Laboratory at Woods Hole, Massachusetts, students can watch the small Gonionemus jellyfish swim rapidly to the top of the water with a few jets, then turn over and sink slowly to the bottom, with tentacles outstretched for prey. These umbrella-shaped animals contract and force a current of water out through an opening on the underside. They rise for an inch or two and then repeat the process. Squids are capable of using this method much more efficiently with their better-developed muscles, streamlined bodies, and a more precise mechanism for directing the jet. However, the same type of jerky movement results, and no animal has been able to solve the problem of how to produce a rapid continuous movement by jet propulsion.

Wormlike movements. Of all the worms the roundworms, or nematodes, probably have the least efficient mode of locomotion. Almost any wild vertebrate has some of these parasitic roundworms in its digestive tract. If we obtain some alive and place them in a dish, they writhe like snakes but make almost no progress unless their tail tips happen to get braced against a solid object. They are covered with a thick, smooth cuticle which is good protection against the digestive juices of their hosts and is so slippery that they are scarcely disturbed at all by the intestinal contractions around them. The same smooth surface makes it difficult for them to move under their own power. Nonparasitic roundworms living in the soil are little more efficient in movement, although some of them have spines and make better progress. Since they have almost no sense organs, the total behavioral adjustment of nematodes is very crude.

Wriggling movements are also used by freely swimming annelid forms such as the nereids of our ocean shores. With a much better muscular system, and appendages which catch the water, they are able to move more rapidly in either swimming or crawling. The same kind of movement is used by snakes while swimming, but wherever employed it remains a relatively slow and inefficient means of progress.

Earthworms move by a kind of peristaltic action. The back part of the body is squeezed together while the front part is extended to a point where a new purchase can be obtained. The rest of the body is then hauled up an inch or two. Similar methods of inching along are used by the flatworms.

All these types of locomotion are relatively slow and laborious. Most of the animals which use them lead sedentary lives in holes, under rocks, and in other protected places, since they cannot escape from danger by rapid movement.

Appendages. In wormlike movements the whole body is used, but many animals have developed specialized organs for locomotion. Perhaps the most peculiar of these organs are the tube feet of echinoderms, which extend by hydraulic pressure, attach themselves by suckers, and then contract by muscles. Hundreds of these organs have to be coordinated, and occasionally a tube foot will cling too long and be left behind. Starfishes cling like glue to their native rocks, but with all this complicated apparatus their progress is extremely slow.

An octopus crawling over rocks uses its arms in much the same way as the starfish uses its tube feet. Although much faster and better coordinated, it still has to throw out its arms, take hold with its suckers, and pull the rest of the body ahead. The general result is slow and clumsy, rather like a man in a boat throwing an anchor ahead and then pulling on the rope.

Other mollusks, such as certain sea snails, have fins on the side of the foot. The sea-hares, with their rabbit-like shapes, almost seem to hop in slow motion as they move through the water with waving fins. However, the mollusk foot is more typically a mass of muscle which is used either for crawling or for digging in mud or sand. The crawling mollusk uses a type of wormlike movement, but in only one part of the body, which drags the rest after it. The result is so gradual that "snail-like" has become a synonym for slowness.

Arthropods have the first really efficient appendages—paired, segmented motor organs, whose forerunners are the bristles and fleshy, paddle-like parapods of annelid worms. Many arthropods have long, jointed legs, with an arrangement of skeleton and muscles which permits the magnification of muscular movement by a series of levers. Motion can be very rapid, as anyone who tries to catch a so-called daddy-longlegs will discover. Some arthropods have an even speedier means of locomotion—the insect wing.

One of the primary appendages of vertebrates is the tail. In the flexible tail fin of fish, a small lateral movement at the base is transformed into a larger pushing movement at the tip, so that a few rapid flips will send a fish several yards at top speed. The same principle is used in the flippers of sea-going mammals such as whales and seals, although those of the latter are really hind limbs transformed into tail-like organs. Other aquatic vertebrates, such as salamanders and alligators, also use the tail for swimming, although in a less efficient fashion. Birds use the tail as a sort of rudder during flight, and in the land mammals the tail is put to such other uses as a fly whisk in cattle or a grasping organ in opossums and monkeys.

The paired appendages of vertebrates are similar in many ways to those of arthropods. The main difference between them and the arthropod limbs is that the muscles are attached to an internal rather than an external skeleton. The chief result is to make possible greater size rather than to improve efficiency. An external skeleton and the necessity for molting makes growth difficult, and a lobster which attains a length of two feet is a whale among arthropods. At one time in geologic history there were dragonflies with a wing span of two feet, but arthropods have long since abandoned competition with vertebrates in the larger sizes. On the other hand, the vertebrate appendages have been limited in their evolutionary modification by the fact that there are never more than two pairs. These have been modified into fins, legs, and wings; but where the limbs acquired a new specialization, they usually have had to give up an old one. In most birds the fore-limbs have been adapted for flight and the hind limbs for walking, but there is no bird with both good wings and good running legs. Such limitations are not so evident in insects, which have six legs in addition to wings and can often move rapidly both on the ground and in the air. The advantages of large numbers of appendages are even more apparent when we consider the grasping and manipulating of objects.

MOTOR CAPACITIES: MASTICATION, PREHENSION,
AND MANIPULATION

Many of the motor organs of animals can be used for food-
getting and for manipulating objects as well as for moving the
whole body. In the higher animals there is a tendency to develop
special organs for each function. Species with special abilities
for grasping and moving objects vastly increase their ability to
adapt to the environment, and an animal which has good manipulatory organs tends to be rated as intelligent. Raccoons, which
have front feet like hands, are usually thought to be much more
intelligent animals than cats, which are unable to grasp objects.
Cats are relatively poor at escaping from boxes with a complicated latch, whereas raccoons are quite good at this sort of thing.
The various kinds of manipulation which animals use are briefly
discussed below.

Cilia. These microscopic "hairs" are useful for the movement
of only extremely minute objects but are nevertheless employed
in the feeding processes of such relatively large animals as clams,
which create currents of water through the gills by means of cilia
and so are able to filter out microorganisms for food. The amount
of selectivity and variability which can be obtained with cilia is
quite small, and in most cases the animals are able only to start
and stop the movement. As mentioned before, cilia are also found
in the lining of the respiratory tract of mammals, where they help
to remove mucus and small particles from the body.

Tube feet. These slowly reacting hydraulic organs are used
by echinoderms for handling and manipulating food. It is easy to
see them in action by feeding a starfish a small clam. It moves the
clam to the mouth region at the base of the arms and slowly pulls
the shell apart with the tube feet. Then the starfish everts its
stomach through its mouth and digests the clam. This is a very
complex process, comparable to the highly coordinated behavior
of higher animals. But it is also extremely slow, and no starfish
would be able to handle a rapidly moving organism.

Tentacles. These can be efficient grasping organs, especially
when they are equipped with suckers as they are in squids and
octopuses. These animals are able to catch and overpower fairly
large and rapidly moving animals like fish and crabs and bring
them to their mouths. Octopuses even use their muscular arms

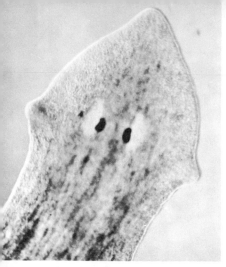 

Sense organs determine the range of contact and stimulation from the environment. The sensitive lobes on the head of a planarian flatworm touch and taste the immediate surroundings. The simple eyes detect only differences of light and shade; they are not image-forming. (Photo by P. S. Tice)

Insect antennas extend the range of stimulation. In the nocturnal Cecropia moth, whose head is shown here, the antennas are used to smell as well as touch. In this way some moths can locate their mates at distances of half a mile. The compound eyes form images but are not adapted for great distances. (Photo by R. B.)

Eyes and ears greatly extend the range of contact. In the tarsier (a primate more primitive than monkeys) the eyes are unusually large and adapted for night vision. The large external ears are an aid to good hearing in an animal that must climb about in trees in the darkness. (Photo by R. Buchsbaum)

Arthropod appendages are, on different segments of the body, structurally specialized for a variety of behaviors. On the underside of the front end of a centipede we see: sensory antennas for investigation, mouth parts for manipulating and ingesting food, poison claws for killing prey, walking legs for locomotion. (Photo by P. S. Tice)

Tube feet of starfish are both locomotor and manipulative. Starfish shown here is using tube feet to cling to glass front of aquarium and to pry open shell of mussel it is about to feed on. Tube feet are fluid-filled and extended by hydraulic pressure; retracted by muscular contraction. At their tips are muscular suckers that attach strongly to surfaces. (Photo by W. K. Fisher)

Numerous arthropod appendages permit a great variety of adaptations. Aquatic insect, the water boatman, with oarlike segmented appendages, can still use its wings to fly. (Photo by P. S. Tice)

Two pairs of vertebrate limbs restrict variety of adaptation. Penguin has wings modified for swimming and is unable to fly, but lives where there are no large land predators. (Photographed in Bristol Zoo, England, by Monte Buchsbaum)

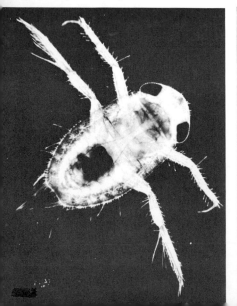

to build nests of small stones. However, tentacles do not contain skeletal elements and cannot make use of leverage. They are difficult to support and extend except in the water, and very few land-living animals make use of them. Even in water-living animals, tentacles can be used efficiently only to draw objects toward the base of the tentacle.

Arthropod appendages. The arthropods have a large number of paired appendages, and these are often specialized for many different purposes in the same animal. In a lobster some of the appendages are used for creating a current of water over the gills, others for walking, and the big claws for grasping and crushing. Appendages around the mouth are used for holding the food in place and chewing it. In insects there is a similar specialization, and the manipulative skill which these animals achieve is probably the finest in the animal kingdom. The nests of paper-building wasps and the honeycombs of bees make any similar structure built by vertebrates look crude by comparison. Even the primates with their well-developed hands cannot achieve this kind of precision. Man has been able to duplicate the feats of insects only with the aid of tools and machines.

Vertebrate limbs. As mentioned above, vertebrates are handicapped in anatomical specialization by their small number of appendages. This problem has been solved in various ways, usually at the expense of one function or another. One device is to use all four appendages for locomotion and the jaws for prehension, as is done by fish, by many herbivorous mammals, and also by some carnivores, like the dog. However, the vertebrate jaw is chiefly adapted for holding and carrying objects and very little fine manipulation can be done with it. Animals which are adapted in this way tend to be relatively poor at manipulating objects and seldom build any kind of complex shelter. The exceptions are the birds with their jaws encased in a finely pointed bill capable of controlled manipulation.

In flying vertebrates such as birds, the forelimbs are used for flight and the hind limbs and mouth can be used for grasping. Orioles show a high degree of manipulative skill as they weave their hanging nests with their beaks. However, feet specialized for clinging and grasping are no longer efficient in running. The perching birds can only hop, and hawks and owls are scarcely able to move at all on the ground. Some of the water-living birds such as ducks have adapted the hind feet for swimming but have lost all power of grasping with these organs.

In the one type of mammals capable of prolonged flight, both appendages are used for flight. Bats retain some ability to grasp in both fore- and hind limbs but use this chiefly in roosting. They are clumsy on the ground and snatch their prey with their mouths while on the wing.

In many land vertebrates the hind limbs are used for locomotion and the forelimbs for grasping and clinging. The frog is an animal of this sort, and some of the extinct dinosaurs apparently ran around on their hind legs like chickens. Mammals of many different groups have a hand on the forelimb. The jumping kangaroos and wallabies of Australia are able to use the front feet for grasping and manipulating. Among the carnivores the raccoons have developed extremely efficient hands. The same tendency is seen in small rodents which use the front feet to hold food or handle nest-building material and can even be taught to pull strings. Squirrels are especially clever at manipulation, as anyone can testify who has watched a red squirrel efficiently dissecting a pine cone with paws and teeth. The kangaroo rat uses the small forelimbs as hands; only the large hind feet are used for running and jumping.

Grasping and manipulation are often associated with the habit of living in trees. A great many of the primates are arboreal and use both front and hind feet for grasping. Those which swing from branch to branch have long, strong fingers; those of man and the gorilla are relatively short. All show some tendency to specialize the forelimbs for superior manipulation. The chief advantage of the primate hand is a thumb which can in many species be opposed to the other digits. This makes greater precision possible in handling small objects. In some of the great apes the big toe is almost as efficient as the thumb. Man is unique among primates in the degree to which the hind limbs are used only for locomotion and the front limbs for grasping and prehension. Other ground-living primates, such as the baboons, use both front and hind limbs for locomotion, with some loss of manipulative ability.

Some arboreal monkeys use the tail for grasping objects. This is one of the exceptions to the general rule that vertebrate adaptations for prehension are limited by having only four limbs. An even larger exception are the elephants, which use all four limbs for locomotion but have adapted the nose as a prehensile and manipulatory organ of considerable skill. While its trunk has some of the defects of tentacles, an elephant is nevertheless able to handle anything from small leaves and nuts to large branches of trees. In common with other animals with good manipulatory ability, elephants have the reputation of being especially intelligent.

We conclude that the ways in which an organism can adapt to its environment are profoundly affected by its motor capacities as well as its sensory abilities. Differences in sensory capacities make it possible for animals to respond to far different ranges of stimuli in the environment. Differences in the power to move allow animals to live in greater or smaller areas and to move or not move from one environment to another. Differences in the ability to grasp efficiently and manipulate objects determine an animal's power to alter and modify its environment. Still another capacity which affects adaptiveness is the ability of the central nervous system to analyze sensory impressions and transmit stimuli to the motor organs.

CAPACITIES OF THE CENTRAL NERVOUS SYSTEM

The form and organization of the nervous system of any animal is always closely related to the kinds of sense and motor organs which it possesses. However, there are certain ways in which the anatomy of the nervous system itself has an effect upon its function. Structure may affect both the speed of nervous action and the amount of general coordination of the body.

Transmission of stimuli. When we watch a cat running away from a dog, it looks like a spinning blur of legs and tail because it moves so fast. On the other hand, if we happen to be at the seashore and watch a large jellyfish swimming through the water, we are impressed with the very slow way in which it moves, lazily contracting its bell every few seconds and advancing almost imperceptibly. This difference in speed is in part the result of different kinds of motor organs, but it is also partly caused by the speed with which nervous stimuli can be transmitted. In the brain of a cat, stimuli may travel as fast as 119 meters per second. Since a cat is less than a meter long, a nervous stimulus traveling at this rate could pass from nose to tail tip in a very small fraction of a second. In contrast, a nerve impulse may pass through a jellyfish as slowly as 0.15 of a meter per second. In a large jellyfish it could take a whole second for a nerve impulse to pass from one side to the other. The rate of nervous activity in these lower animals is roughly one thousand times slower than that in a mammal.

Why should the rate be so much slower? One answer is temperature. In a cold-blooded animal all reactions are slowed down when the body temperature drops to correspond with a lower outside temperature. Snakes and lizards are torpid in cold weather, although they move very quickly in the hot sun. Water-living animals like frogs reflect the low temperature of the water around them. The fastest speeds recorded in frog nerves at 21.5° C. are not half those in warm-blooded animals.

Another thing which affects the speed of nervous action is the thickness of insulating material around the nerve. The fastest nerves of a cat are covered with the heaviest myelin sheaths. Some nerves with a thinner sheath are much slower, and the slowest are the sympathetic nerves, which have almost no sheath at all. Their speed is only 1 or 2 meters per second, which means that reactions of the internal organs are always slower than those of the skeletal muscles. Almost all invertebrate animals have nerves with thin sheaths and hence slow nerve impulses. One of the fastest reactions recorded for an invertebrate is in the leg nerve of a lobster, where impulses travel at the rate of 12 meters per second.

A third factor which slows down a nervous stimulus is the number of nerve cells through which it has to pass. At each connection between two nerve cells, or synapse, the impulse is slowed down. One long single fiber is much faster than several short ones. In the sea anemone Metridium, there is a network of many separate nerve cells and the rate of transmission is only 0.12 meters per second. It also has a few direct tracts of longer cells through which a stimulus goes much faster.

The nervous system of invertebrate animals is therefore basically much slower than that of vertebrates. However, many invertebrates move quite quickly in an emergency, as when a crayfish darts back with a sudden flip of its tail. This ability is the result of a special structure. A stimulus goes faster if the diameter of the nerve fiber is larger. Many animals, like the crayfish, have special giant fibers which are formed of several nerve cells fused together. In the squid these fibers go to the muscles of the mantle. In an emergency the squid suddenly contracts these muscles, shoots out a stream of water, and rapidly retreats. Even such slow-moving animals as earthworms have giant fibers running down the central nerve cord which enable them to pull back into their holes in a hurry when startled by sudden light or touch. However, even the giant fibers are not nearly as fast as the thickly sheathed nerves of warm-blooded vertebrates.

The physiology of stimulation. The nerve impulse is a very rapid chemical reaction which involves a change of electrical polarity in the surface of the nerve fiber and from which there is a very quick recovery. Some of the most interesting and important principles affecting behavior can be illustrated by simple laboratory experiments on an isolated nerve and muscle of a frog. Direct stimulation of the nerve by electrical shock shows that it reacts in accordance with the "all or none" law; that is, a single nerve fiber either reacts completely or does not react at all. Too small a stimulus gives no reaction, but any stimulus above the effective point results in the same complete reaction.

Accommodation. We can derive another general principle of nervous physiology from the fact that the effect of an electrical shock is dependent on the rate of change. Stimulation by direct electric current occurs only when the current is turned on or off. When we apply a very weak current to a nerve, nothing happens. If we increase the voltage very slowly, there is still no response, and the nerve is said to accommodate itself. However, if we increase the voltage to the same point more suddenly, the nerve will respond. Thus we see that the general behavioral principle— that stimulation consists of change—has a basis in the physiological nature of nervous tissue. Presumably this also accounts for the phenomenon of accommodation in sense organs which was described in chapter 1.

Summation. Another physiological principle which is directly involved in behavior is that of summation. An electrical stimulus which is ordinarily too weak to produce a reaction in a nerve can, if repeated very rapidly, cause the same effect as a single strong stimulation. The question now arises, Does the principle of summation apply to the behavior of the whole organism as it does to individual nervous tissues?

The answer appears to be Yes. An animal which is stimulated

in one respect tends to be more reactive to other stimuli. If hungry, it frequently responds to slight stimuli which would not affect it otherwise. It may thus exhibit behavior which has no relation to the problem of getting food. Hungry animals often show an increased tendency to fight. An animal primarily stimulated to fight may also show the effects of summation in its reaction to other stimuli. The male stickleback fish which is defending its territory against another but is not actually approached so that a fight can start may become highly stimulated and begin digging a hole in the sand as if it were making a nest. In this "displacement" behavior, the stickleback is presumably responding to slight nest-building stimuli which would not ordinarily affect it.

In conclusion it may be said that the study of the physiology of stimulation gives a sound basis for the general principle that stimulation consists of change. Rapidly repeated or summatory stimuli produce a more abrupt change than the slowly increased stimulation which produces accommodation. The effect of each can be explained by the constant rate at which the nerve recovers from stimulation. However, neither of these principles will account for prolonged reactions after stimulation has ceased. As we shall see in the next chapter, there are brain centers which may have the special function of prolonging and magnifying stimulation.

Concentration of nervous control. Most many-celled animals have some concentration of nervous tissue. Instead of following separate pathways, sensory nerves tend to travel to a central area from which the motor nerves emerge. There are three principal ways in which such a central nervous system may be constructed. Coelenterates and echinoderms have a radially symmetrical body with a ring of nerve fibers. In the jellyfish this ring runs around the margin of the bell near the bases of the tentacles, and in the starfish the most important ring of nervous tissue encircles the mouth. In both cases there is relatively poor motor coordination. A starfish turned on its back rights itself very slowly, at first being active with all five arms. Finally two of the arms take hold and flip the animal over. Being nearly alike in all directions, the starfish constantly faces the problem of which stimulated region is going to control all the rest.

A second type of central nervous system is found in mollusks. Paired masses of nerve tissue, called ganglia, are developed in

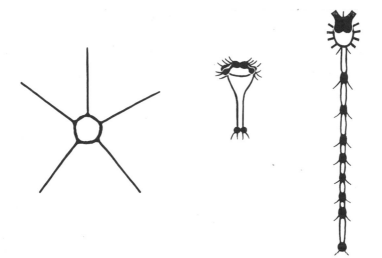

Fig. 7. Three basic types of nervous systems:
Left.—The starfish has a simple nerve ring around the mouth, with a radial nerve leading off to each arm. This sort of nervous system is typical of animals whose bodies are radially symmetrical. Their movements are relatively slow and poorly coordinated.

Center.—Mollusks have an unusual type of nervous system in which each main part of the body has its own pair of nerve centers, or ganglia, one for each side. The three main ganglia are connected with each other in a triangle on each side of the body. In an active mollusk, like the free-swimming sea slug *Aplysia* illustrated here, the ganglia are concentrated in the region of the head, with only the visceral ganglia at a distance.

Right.—The scorpion has a heavy concentration of nervous tissue in the head. This "brain" is connected with a longitudinal nerve cord having a ganglion in each section of the body. A large number of animals have central nervous systems of this general type. That of vertebrates is essentially similar except that there is greater concentration of nervous tissue in the brain and spinal cord and no ring around the mouth (see p. 56). (Original)

several parts of the body: the head, the foot, the viscera, and the mantle. Motor coordination is often inferior, but this may be the result of extremely simple motor apparatus. Squids and other cephalopods, with their better motor organs and with concentration of most of the pairs of ganglia in the head, are capable of highly coordinated activities.

Finally, there are several types of nervous systems composed of longitudinal nerve cords with greatly enlarged ganglia, or brains, at the front end of the body. These reach their highest development in arthropods and vertebrates, both of which primitively show a segmental organization. Insects and other arthropods have a brain but retain separate segmental ganglia. In the vertebrates the segments of the central nervous system are entirely fused, and the only enlargement is the brain at the anterior end. Anatomically at least, the vertebrates have a more centralized organization.

The concentration of nervous tissue in the form of a brain is always closely connected with the sense organs. In a large-brained insect like the honeybee, there are special brain lobes connected with the eyes, and if these are removed, the animal can no longer

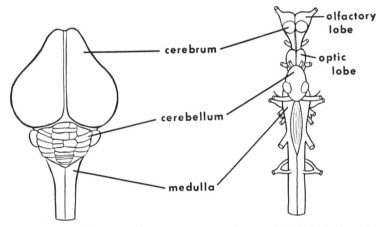

Fig. 8. Brains of two vertebrates, guinea pig, *left,* and dogfish shark, *right,* drawn so that the spinal cords are approximately the same size. By comparison, the cerebrum of the guinea pig is immensely enlarged. The development of this "neocortex" of the cerebrum makes possible much more complicated and variable types of motor control and learning. (Original)

see. Another general characteristic of brains is that they control relatively elaborate patterns of behavior. In the hermit crab the removal of the brain does not affect ordinary locomotion, which is controlled largely by the segmental ganglia and their connections, but a brainless crab removed from its snail-shell shelter is unable to find its way back into it without assistance.

The same general characteristics are found in the brains of vertebrates. In a shark almost the entire anterior part of the brain is composed of the olfactory lobes, which are connected with the sense organs through which water is tasted. In the higher fishes and birds, the optic lobes are larger, corresponding to the importance of vision in these animals. As we examine the brains of reptiles and mammals, we find that there is a greater enlargement of the front part of the brain, culminating in the neocortex of mammals that forms the greater part of the cerebral hemispheres. This new part of the brain has two general functions, which can be demonstrated by experiment. It receives stimuli from other sense organs, particularly the eyes; and injury to it may cause blindness. It also controls all sorts of complicated motor responses, as can be shown by electrical stimulation. Like the hermit crab mentioned earlier, a mammal which has had large areas of the forepart of the brain removed is incapable of any complex patterns of behavior.

Enlargement of the brain greatly increases the number of nerve cells, so that the possibility of different sorts of nerve connections is greatly magnified. Along with this a greater variety of behavior is possible and also a greater degree of recovery from injury. A rat whose brain has been injured may at first appear blind and lack some of its motor behavior, but after a few weeks it will largely recover. If enough of the brain is left, the function can be transferred from one part to another. It also appears that species with a larger development of the neocortex are capable of more complicated kinds of learning.

Capacities for the organization of behavior. The simplest sort of behavioral organization is that in which a specific stimulus invariably produces a specific response. In this case the behavior of the individual is organized by its hereditary constitution. Instances of this kind are actually very difficult to find in living organisms, except in immature animals or where parts of the body have been surgically isolated. In most animals there is at least some modifiability of behavior, even in the protozoans.

The capacity for organizing behavior is severely limited in the lower animals by restrictions on their sensory and motor activities. Even if an animal like a planaria has the power to organize its behavior, it is very difficult to demonstrate it. In many species motor reactions are so fixed by heredity that behavioral organization can be studied only as the animal travels from one place to another. This accounts for the popularity of the maze technique in studying problem-solving ability.

Among the higher animals also, a species which has limited sensory or motor capacities has limited powers of behavioral organization. It is of no use to present a dog with two sections of stick which can be fitted together to reach a bit of food. The dog simply does not have the motor apparatus necessary for this task, although chimpanzees can do it successfully. An experiment which takes advantage of an animal's natural capacities must be done before any judgment can be made.

We should be extremely cautious and conservative about concluding that any species of animal has or does not have the basic capacities for organizing its behavior, until the right kind of experiments are done under conditions favorable to its abilities. Much of the evidence on comparative intelligence of animals is suspect on these grounds. The most that can be said is that there is some evidence of differences in this capacity. For example, most invertebrates seem to have to repeat an experience dozens of times before an association is formed, in contrast to the "one-trial learning" of birds and mammals.

On the other hand, the octopus, which has highly developed eyes, can form associations in much the same way and with the same ease as do the so-called higher animals. A small variety of octopus lives on the sea bottom in the Bay of Naples and can be easily caught and kept in the aquariums of the zoological station nearby. Two visiting British scientists, Boycott and Young, decided to test the learning abilities of these animals. They placed a few bricks in one end of an aquarium, and the resident octopus immediately retired behind them. Then the experimenters dangled a crab by a thread at the other end of the tank. The octopus has large pigment cells in the skin which can be expanded or contracted by nervous control. As soon as it saw the crab, the octopus turned dark, moved swiftly down the tank, captured the crab, and returned to its pile of bricks. Next the experimenters dangled a crab with an electrified white plate behind it. The octopus attacked the crab and was shocked. At this point it turned pale and retired hastily to its bricks, blowing jets of water across its arms and wiping them on the mouth region. Its next attacks were a great deal more cautious and hesitant, and after a few trials it made no more mistakes. All the animals tested learned as quickly. They simply leaned out from their bricks, took a good look to see if the crab had a white plate behind it, and then attacked or not, according to the circumstances. In other words, they learned to discriminate between crabs that came alone and crabs

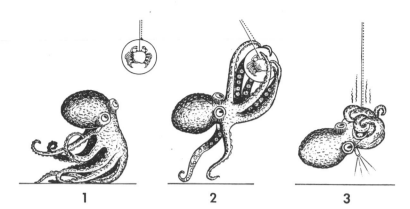

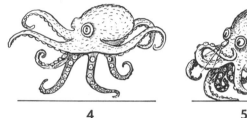

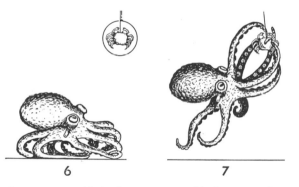

Fig. 9. An octopus, with its large eyes, readily learns to distinguish between painful and harmless objects. (1) A crab attached to a white plate is
lowered into the water. (2) The octopus approaches and seizes the crab.
(3) The electric current is turned on. (4) The octopus retreats, turning pale.
(5) The octopus blows water over the tentacles which have been shocked.
(6) When the crab on the white plate appears again, the octopus avoids it.
(7) It seizes and eats the crab when it is presented without the plate.
(From sketches by Vivian S. Hixon of an experiment described by Boycott
and Young)

that had a plate behind them. As we shall see, there is very little difference between this performance and that of dogs in a similar learning situation. We may cautiously conclude that the capacity for organizing behavior is strongly correlated with sensory and motor capacities.

RELATIONSHIP BETWEEN BASIC CAPACITIES AND BEHAVIORAL SYSTEMS

Most of us have seen the long-armed gibbons swinging gracefully from trees and bars in a modern zoo. This motor specialization in part determines the patterns of agonistic behavior of which these animals are capable. In their native haunts they can escape rapidly through the trees, but on the ground they are almost helpless, being able only to run awkwardly for short distances, like a man on crutches. In the same way, every other species has its own special sensory-motor abilities and limitations. By summarizing these facts, we can draw certain general conclusions about the relationships between basic capacities and modes of adaptation.

Consequences of sensory differences. The sensory abilities of an animal chiefly determine its power to discriminate between different parts of the surrounding physical and biological environment. Tactile and chemical sense organs reach only short distances, and there has been a trend in evolution toward greater development of the sense organs of sound and sight, which permit contact with the environment at greater and greater distances.

Sensory ability, of course, has its largest effect upon systems of investigative behavior. Animals which have only tactile senses must continually "feel out" their environment in much the same way in which a paramecium continually alternates between going ahead and backing up, depending on what it strikes. Nocturnal rodents such as mice and rats use the tactile senses a great deal, and this is what gives a mouse its air of constant scurrying activity. By contrast a bird, which depends on sight, will often limit its investigative behavior to simply turning its head. That animals with good eyes actually spend a great deal of their time inspecting their surroundings is shown by Butler's experiment. He placed one monkey at a time inside a cage with solid walls and a covered window. This window could be opened by the animal but closed itself automatically after a few seconds. The monkeys kept opening the window for hours at a time to see what was going on outside. Some things interested them more than others. They would open the window much more frequently to watch a moving electric train than they would for stationary groups of blocks. As might be expected, the thing which fascinated them most was another monkey.

The allelomimetic behavior seen in schooling fishes and flocks of birds is likewise strongly affected by sensory capacities, since the animals concerned must be able to keep track of and follow each other's movements, sometimes at a considerable distance. Consequently allelomimetic behavior is seldom found in invertebrates. Two of the few exceptions are the squids, which have highly developed eyes and swim around in schools like fish, and the army ants, which move in columns by literally "keeping in touch" with each other.

Vertebrates with their superior eyes develop allelomimetic behavior more frequently. This behavioral system is found in numerous species of fish that live in open water, such as the sea-going herring and mackerel, and can be commonly seen in minnows and young fish feeding along the shore of any clear fresh water lake. It is rarely if ever found in amphibia or reptiles but is frequent in birds and mammals. Interestingly enough, related species sometimes differ widely. The elk of the western plains move together in large herds and show a great deal more of this behavior than do the white-tailed deer, which live in small scattered groups in brush and thickets. Animals with the requisite capacities do not necessarily develop allelomimetic behavior, particularly if they are nocturnal or their habitat contains many obstructions to vision.

Since allelomimetic behavior is so greatly dependent on the sensory capacities which enable one animal to observe another closely and on the motor capacities to copy its activity, the best examples occur in animals with good eyesight. Flocking birds on the wing can follow one another's movements with such precision that the flock seems to act like a single individual.

To summarize, differences in sensory capacities have important effects on the distance and accuracy with which animals are able to respond to stimuli. Patterns of investigative behavior are strongly modified, and this in turn affects the ability of animals to keep in touch with one another and coordinate their activities in allelomimetic behavior.

Finally, the ability of animals to discriminate between members of their own species has a considerable effect on social behavior and organization in general, as will be shown in a later chapter.

Motor capacities and adaptation. Any animal which can move at all and can discriminate between different parts of its environment will show some shelter-seeking behavior. Sexual behavior, on the other hand, requires fairly good motor abilities. For the slow-moving jellyfishes it is apparently just as efficient to release sperm into the water and allow them to find the eggs as it would be for the jellyfishes to try to make direct contact with each other. Among mollusks, fresh-water clams discharge the sperm into the water with little overt behavior, whereas most snails, which move swiftly by contrast and can partially emerge from their shells, show true mating behavior as they come together to fertilize their eggs. Animals with better motor abilities, like arthropods and vertebrates, often develop highly elaborate patterns of courtship and mating.

Agonistic behavior is strongly affected by motor ability. Its origins are seen in escape behavior from predators, which depends on some power of movement. A paramecium quickly withdraws from an injury, and even the sluggish ameba slowly crawls away. Sponges and coelenterates and some higher animals such as barnacles do not have the necessary motor apparatus for rapid escape. However, those forms which can move at all retreat or withdraw in some way. Even clams can disappear quite rapidly into their native mud, as anyone who tries to dig them out soon discovers. Snails, turtles, and other animals with hard shells often escape by simply withdrawing into their armor. This comes very close to the passive reaction to injury which is developed to its greatest extent in the death-feigning of some higher animals. An opossum which is overpowered will go completely limp and apparently lifeless for several minutes, then suddenly bound to its feet and escape if it is no longer held. Similar reactions are seen in turkey buzzards.

The capacity for inflicting injury upon an animal of another species is found in a defensive form in the coelenterates, which use the stinging capsules on their tentacles to defend themselves against injury as well as to capture prey. How effective these can be is demonstrated by the Portuguese man-of-war, whose stings have occasionally proved fatal to bathers.

It is sometimes difficult to distinguish between agonistic behavior and the capture of prey. True offensive fighting, in which one animal actually attacks another of the same species, is scarcely seen at all except in those arthropods and vertebrates which are capable of rapid movement. Anyone who has caught crayfish in the spring and brought them into the laboratory will realize that arthropods can fight. The males scuttle around the aquarium attempting to turn other crayfish on their backs in the mating position. If two males try this with each other, a struggle results, ending only when one crunches off part of the other's legs or antennas.

In the vertebrates fighting is almost a general characteristic. Nearly all species of mammals and birds show some kind of fighting, and conflict is common in fish and reptiles. Besides speed, a fighting animal usually needs some kind of structure such as teeth, horns, or claws with which it can inflict damage on others, but this is not essential. Among the peaceful amphibians, which possess neither teeth nor claws, there is a species of frog which defends its territory by jumping heavily on the backs of intruders. Nor does the ability to fight necessarily imply that fighting occurs. Howling monkeys have reduced fighting to loud and belligerent howls from a distance. All of this leads us to conclude that agonistic behavior is made possible, though not inevitable, by motor organs which permit escape or attack.

In vertebrates there is a tendency to evolve special patterns of agonistic behavior that are used only in conflicts between members of the same species. Among fish, an animal may dart toward another with essentially the same movement that it might use to secure a floating worm, but even these animals may evolve special postures of the fins and different colors which are only used against a member of the same species. Among predacious mammals quite different patterns are employed in attacking a prey animal than when attacking a member of the same species. Even the prey animals may make the same sort of distinction. For example, a bull moose pushes with its antlers when attacking another moose but strikes with its forefeet when resisting the attacks of a pack of wolves.

A second important capacity of motor organs is the ability to grasp or manipulate objects. This extends the range of adaptability in many types of behavior. A paramecium, which may take both dirt and food material into its food vacuole, contrasts greatly with the sea anemone, which selects pieces of meat with its tentacles and pushes them into its mouth. A still more gifted animal like the grasshopper may grasp the stalk of a food plant with its legs while the appendages of the head rasp the food away, chew it, and push it into the mouth. Even more complex activities are the food-gathering of bees and the food-storing of rodents like the ground squirrels, which transport food in their cheek pouches and bury it in particular spots.

Epimeletic or care-giving behavior is strongly affected by manipulative ability. Many species of animals with little manipulative ability have some sort of brood pouch in which the care given the young is entirely a matter of physiological function and involves no behavior. Such pouches may occur in animals as tiny as rotifers or as large as clams.

It is only in arthropods and vertebrates that we find care of the young developed to a high degree. Many arthropods give some protection to the young. A female lobster ready to lay eggs lies down on its back and curls its abdomen forward in such a way that the emerging eggs become attached to the swimmerets; there they remain until the larvae are sufficiently developed to swim off by themselves. Spiders spin nests for their developing young but provide no food for them. The great development of epimeletic behavior occurs in the social insects. Species such as the paper-nest wasps build a highly elaborate nest in which the young are protected, handled, cleaned, and fed.

Care-giving behavior often has the function of substituting the protective behavior of the adult for adaptive behavior of the helpless young. Substitution for ingestive behavior is common, but the species must be able to manipulate food in order to bring it to the young. Fish, which are very poorly provided with manipulative organs, never bring food to their young, nor do hoofed mammals like the deer. All mammals have bypassed the need to handle food by the device of feeding the young through mammary glands, but additional feeding with solid food is common in mammals which can manipulate. Another kind of care-giving behavior is cleaning the young by removing excreta or by grooming. In primates this is extended to animals of all ages. A common sight among chimpanzees is mutual grooming, which depends upon the ability to pick small objects out of the hair.

We conclude that motor abilities are basic for all types of behavioral adaptation. The capacity to move effectively is necessary in both sexual and agonistic behavior. Manipulative ability is particularly important in ingestive behavior and in the care of others. A final important way in which motor capacities affect behavior is in the development of communication, which will be discussed in a later chapter. Most communication is accomplished through signals, and the transmission of complicated messages requires elaborate motor apparatus.

Capacities of the central nervous system. We have seen above that the capacity to organize behavior is strongly correlated with the possession of adequate sensory and motor apparatus. Apart from this, what effect do different nervous capacities have upon behavior? In the first place, there are large differences in the speed of reaction. This means that highly organized animals can adapt much more swiftly than the lower ones and often have a great advantage over them in a competitive existence. With their swift reactions, fish have almost entirely replaced the slower-moving cephalopods as the dominant water-living animals. In the second place, nervous capacities alter behavior by permitting more complicated organization. With a highly developed brain, a mammal is able to associate sensory stimuli and motor reactions in many different ways and to coordinate responses like manipulation into highly complex activities. The possession of a large brain permits a greater degree of escape from the hereditary organization of behavior. In the higher animals behavior tends to be organized chiefly on a basis of learning and experience, as we shall see in later chapters.

The differential capacities of animals involve the study of function or physiology. Unlike anatomy, which merely limits the possibilities of adaptation, physiology is a part of behavior itself. More than this, physiological processes may themselves act as stimuli. These physiological causes of behavior are the subject of the next chapter.

CHAPTER 4

Internal Causes
The Physiology of Behavior

A major problem of the science of animal behavior is what makes behavior occur, and the simplest answer is provided by the stimulus-response theory. If behavior is adaptation to change, then it should always be caused by some change or stimulus in the environment. But when we apply this theory to particular cases, it does not always seem to fit. Anyone who has ever kept poultry has encountered the problem of the broody hen. For months the average hen is a calm and placid creature, producing an egg every day or two with no more than a brief cackle. She runs for the feed trough when you appear with grain, runs away squawking when you try to catch her, and generally leads a simple round of predictable activities. Suddenly, with no apparent change in the environment, all this is changed. The hen loses her interest in food and sits continually on the nest. When you touch her, she no longer runs away but ruffles up her feathers and pecks savagely. She stops laying eggs and spends practically all her time sitting on them instead. Removing outside stimulation makes little difference. If you take the eggs away, she will brood on the empty nest; and if you take away the nest, she will sit on the bare floor or any small round object remotely resembling an egg. Obviously something is affecting her behavior which has little to do with the outside environment.

We need not, however, abandon the stimulus-response theory. Changes can occur within the body of an animal as well as in the outside environment. In the case of the broody hen, her behavior is related to the secretion of hormones. In many other cases the exact nature of these internal changes is either obscure or extremely complex, and scientists have lumped these unknowns under such terms as "motivation," "drive," and "incentive." However, when any case is studied in detail, stimulation can always be traced back to a set of specific physiological changes.

The internal causes of behavior have been thoroughly studied in connection with only a few major types of behavior and within a relatively few species. One of the most obvious and best-known examples of internal stimulation has to do with eating.

INGESTIVE BEHAVIOR

Taking in food is a universal and necessary activity for all forms of animal life and has probably been more thoroughly studied in more species of animals than any other type of behavior. Furthermore, it is convenient for laboratory study because, no matter how strange the environment may be, an animal has to eat in order to stay alive.

Internal changes which affect eating occur even in animals relatively low in organization. Among coelenterates we can easily observe the feeding behavior of the common sea anemone Metridium. The feathery tentacles of this sea "flower" are often seen in protected spots along the New England coast at low tide. Instead of having a thin stalk, as a flower should, the rest of the animal looks like the brown trunk of a tree. When brought into the laboratory, it usually contracts into a small brown ball, but after a while it relaxes and unfolds its tentacles. These have cilia which beat toward the tips so that sand grains and other inedible particles dropped upon them are automatically carried away. When, however, we drop a small piece of crab meat or other delicacy upon the tentacles, they immediately contract and carry it to the mouth in the center of the oral disk, where it disappears. Any small animal which touches the tentacles meets a similar fate.

The tentacles of Metridium will continue to carry food to the mouth as long as we offer it, until the animal becomes stuffed and no more can be swallowed. In many cases this is apparently more than the animal can digest and afterward the whole mass of food is thrown out again. In Metridium the feeding reaction seems to be automatic and is limited only by the body cavity becoming completely full of food.

Other sea anemones are less voracious, and after they have taken a few bits of food, their reactions become slow and finally stop. It looks as if in these animals there is some sort of causal mechanism, which may be labeled "hunger," since it is satisfied by food. Similar reactions are seen in the fresh-water hydra, which reacts readily to food when the digestive cavity is empty. A starved hydra becomes hypersensitive, and a solution of meat juice alone will cause a swallowing reaction. Once fed, it becomes indifferent to further stimulation.

Effects of stomach contractions. The exact chemical and physical changes which take place in the bodies of these simple animals have not been studied in great detail, and much of our knowledge comes from the higher animals. One of the classical experiments along these lines was done by Cannon on a human subject. He asked an assistant to swallow a rubber balloon attached to a small tube connected with a recording instrument out of his sight.

Anyone who tries to repeat this experiment will find that it is not as easy as it sounds. A balloon is quite difficult to swallow and the tube inside the throat produces a very unpleasant sensation. In addition, the stomach juices tend to digest ordinary rubber and unless the balloon is made of special quality rubber, it may collapse with a pop at a critical point in the experiment. Once the apparatus is in place, the balloon is blown up and the other end of the tube connected with a rubber diaphragm which moves a needle and makes a written record. Any pressure on the balloon is registered by the needle.

As we watch the record, we immediately notice the changes in pressure produced by breathing, which indicate that the mixing and churning of food in the stomach is considerably helped by the diaphragm as well as by the action of the stomach itself. In addition to these regular up-and-down movements, the needle shows long and gradual fluctuations in pressure which come from the stomach contractions.

In Cannon's experiment he asked the assistant to press a button each time he felt hungry. Later, looking at the record, he found that this sensation had always been preceded by the contraction of the stomach. He concluded that hunger pangs are caused by the violent contractions of an empty stomach. Thus we have a clue to the nature of one of the internal changes which may affect ingestive behavior. Whether or not this same mechanism exists in other animals is extremely difficult to find out, even though there are means by which balloons can be introduced into their stomachs. In an animal which cannot talk it is difficult to tell whether or not any sensation is being produced. We might also wonder whether the same mechanism is present in ruminant mammals, which have the stomach separated into several different parts. The hunger contractions of the stomach must in turn be caused by something else. The problem is not a simple one, since these contractions, like other reflexes, can be affected by learning. In human beings they are more likely to appear about the time of regular meals than at other times when the stomach is equally empty. The contractions of the stomach are also affected by the level of blood sugar, which fluctuates according to the state of nutrition.

Fluctuations in blood sugar. Experiments on rabbits and rats show how blood sugar may affect behavior. In modern hospitals patients are often fed by introducing glucose and other solutions slowly into the veins, and a similar apparatus can be set up for a rabbit. The animal is confined in a special experimental box with several plates on the floor. If it sits on one, nothing happens. Sitting on another causes glucose to flow into its veins, whereas still another produces a flow of physiologically balanced salt solution. The evidence shows that the hungry rabbit prefers the sensation of glucose flowing into the veins and chooses its position accordingly. Presumably this has some effect on hunger contractions, but there is also the possibility that it makes the animal feel better by a direct effect on the tissues. At any rate this is experimental evidence for a second kind of internal stimulation which affects eating behavior in mammals.

The ultimate cause of internal stimulation appears to be the process of metabolism, which uses up food materials in the blood. The complete process and its effects run something like this: A rat is kept in a bare cage, or perhaps an activity wheel, with plenty of water but no food. After a few hours its activity begins to increase. An active wild rat under natural conditions would move around until it eventually found food. The laboratory rat simply runs around and around in its wheel. In both animals metabolic processes have used up the supply of blood sugar, and this has produced stimulation by either acting directly on the nervous system or increasing the contractions of the stomach. The rat is then given food, and as a result of the process of digestion the blood-sugar level is restored, the stimulation is removed, and activity falls to a lower level.

We may wonder how the internal stimulation works effectively, considering the general principle that stimulation is change. Once the blood-sugar level falls, there should be no further stimulation. This problem may account for the phenomenon of hunger contractions in the stomach, which do occur irregularly and are not continuous and hence could act as stimulation over a long period of time.

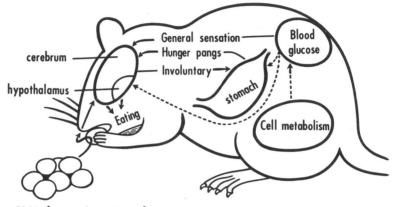

FOOD (External stimulation)

Fig. 10. Physiological mechanisms concerned with the ingestive behavior of a mammal. Dotted lines indicate biochemical or hormonal effects; solid lines indicate nervous stimulation. Note that one source of stimulation is cell metabolism, which is independent of direct external stimulation. This diagram applies only to the ingestion of foods from which glucose is obtained, and a complete diagram of all causes of ingestive behavior would be much more complicated. (Original)

Underlying ingestive behavior are chains of causes combined in a network. Many of these can be traced back to the metabolic and physiological activities which are necessary for life. Some such mechanisms probably exist even in lowly organized animals. This kind of stimulation is to a large extent independent of external conditions, although indirectly influenced by the availability of food. Likewise, it is very closely related to the homeostatic, or automatically self-regulating, physiological processes of the body and is regulated in the same way. The primary effect of glucose takes place in one of the lower brain centers, a special area of the hypothalamus which acts as a "glucostat." Rats whose brains have been surgically injured in this area go on eating without stopping, with the result that they quickly become fat to the point of obesity. The same effect can be obtained by injecting a chemical, gold thioglucose, with which the brain reacts as it would with normal glucose. The gold compound, however, poisons the glucose area so that it no longer functions. Thus eating is controlled by the central nervous system very much in the same way as is drinking, but by a different brain center. It is also possible that changes in the glucose level may directly stimulate sensory nerves.

Special hunger mechanisms. Experiments with the rat indicate that there are physiological mechanisms connected with food elements other than glucose. For example, if a rat is fed on a salt-free diet, it will choose water containing salt rather than pure water. A salt hunger has long been known in domesticated ruminant animals such as cows and sheep, and in wild animals such as deer, which use natural salt licks. The normal physiological balance between sodium chloride and potassium chloride in the blood is not supplied by a herbivorous diet. Vegetable food contains plenty of potassium but little sodium, and the deficiency of the latter has to be made up by eating it.

Encouraged by these findings, nutritionists have attempted to discover whether there are special physiological mechanisms connected with other essential food elements, such as proteins and vitamins. This kind of experiment must be done very carefully to allow for all the factors which may affect an animal's behavior. A rat is given a row of food dishes exactly alike except that each contains a separate purified food. It usually eats more of some kinds than others. However, some rats have a habit of eating only from the end dish and others have a habit of eating from all in succession. Food dishes must be rotated in random fashion. Then too, there are the problems of consistency and palatability, which are difficult to control in purified foods. The general conclusion is that laboratory rats can learn to distinguish foods which contain vitamins from those which do not, presumably because they feel better after eating the vitamin-containing food. They do not, however, show any mechanism regulating the intake of protein. On a diet where protein is separated from other foods, many rats are dwarfed and stunted. It may be concluded that special mechanisms exist for controlling food intake of substances which are likely to be missing from natural foods, but not for all necessary food components. Such experiments have been done almost entirely on laboratory rats and a few farm animals, and it is possible that the causal mechanisms may differ somewhat from species to species. This is especially likely in thirst, which we may consider as one of the special internal changes which affect ingestive behavior.

When the amount of water in the blood falls below a certain level, the flow of saliva is decreased, producing a sensation of dryness in the throat, and this acts as a stimulus for drinking. Internally, there is a controlling center in the hypothalamus that responds to an increase in salt concentration, such as normally occurs as water is used by the tissues and disappears from the blood. Electrical stimulation of this area produces compulsive drinking and also causes the release of an antidiuretic hormone from the pituitary. This acts on the kidney cells to produce a lessened flow of urine and so conserves water. The same hormone is sometimes released as a result of emotional excitement.

These mechanisms exist in many mammals that normally drink large amounts of water, but the situation is probably different in desert-living animals. Some small desert rodents can go indefinitely without drinking. Like other animals they oxidize carbohydrate foods to carbon dioxide and water. This "water of metabolism" is sufficient for their needs. In addition, living in burrows and nocturnal activity has the effect of conserving moisture. When captured and offered water, they refuse it. There are differences even in the common laboratory animals. If a rat is made both hungry and thirsty and given a choice of food and water, it almost invariably chooses the water first and then goes on to food. A guinea pig, whose wild ancestors came from the dry plateaus of the Andes, chooses food first.

As stated at the beginning of this section, ingestive behavior has been more thoroughly studied with regard to its internal causes than has any other type of behavior. We have just seen something of the complex network of causation which determines this apparently simple function. One outstanding generalization can be made from it: *part of the stimulation results from internal changes which are the result of ordinary metabolic processes.* This explains why an animal which is kept in a uniform environment with almost no external stimulation will still show periodic bursts of activity.

This brings up the question of whether the same sort of physiological mechanisms lie behind other types of behavior. Are there "hungers" for fighting and sexual behavior as there are for eating?

AGONISTIC BEHAVIOR

As defined in an earlier chapter, agonistic behavior includes the adaptive responses which animals give when they are either inflicting injury or in danger of being injured. The usual patterns of activity include aggressive and defensive fighting, escape behavior, and passivity.

When we attempt to collect detailed information concerning the causes of this kind of behavior, we find that many facts are missing. Just as most of the details of ingestive behavior have been pieced together from experiments done on man and the rat, a great deal of our information on agonistic behavior comes from two other species. Internal physiological causes have been studied in great detail in the cat and, interestingly enough, the external causes of fighting have been extensively observed in the mouse.

Development of agonistic behavior. If we watch the development of behavior in young mice, we find that the newborn animal is almost completely helpless. It is blind, naked, and toothless; yet if we lightly pinch its tail, it will squeak, move its legs rapidly, and move a few inches away. Escape behavior thus appears very early in life. Later, when the first teeth appear, a young mouse will attempt to bite anything which injures it, marking the appearance of defensive fighting. Still later, when the eyes open, it will take up a defensive attitude with the paws raised when threatened with a moving object. All of this occurs by the time the mouse is twelve days of age. But it is not until the animal is more than a month old that a male will sometimes actually attack another mouse, and then only a stranger.

Here we find one great difference between fighting and ingestive behavior. Mice left without food for a few hours become increasingly active, while mice which have no opportunity to fight show no evidence of internal stimulation but remain peaceable. Males and females almost never fight each other, even when living in the same cage for months; and even a group of males may live without fighting for long periods if they are raised together from birth.

When two strange thirty-five-day-old males are put together, they first investigate each other cautiously with their noses. One of them may start to groom the other, sometimes getting rougher and rougher. The groomed mouse tries to throw the attacker off, and a fight starts. In another pair, one male may attempt to mount the other in sexual fashion, with a similar result. Once a fight has begun, the two males roll over and over, kicking and biting each other. As soon as one is able to hurt the other, the injured mouse begins to run away with the attacker on his heels. If the defeated mouse is unable to escape, he may stop and show the defensive posture with paws outstretched in a helpless way. If completely unable to escape, a mouse may lie flat on the floor and passively submit to the attacks in a way which reminds us of the death-feigning of other animals.

The primary stimuli which set off fighting between male mice are somewhat variable and indefinite. Odor has some effect, as a pair of mice are more likely to fight in the presence of the odor of a strange male. However, the odor of a familiar male does not stimulate fighting. A common clue seems to lie in pain. While it may be difficult to get two mice to fight under many circumstances, an inexperienced mouse always fights back if he is attacked by another. We can conclude that in this species the primary stimulus which brings out agonistic behavior is pain, that pain in moderate amounts will cause the animal to fight back, and that in excessive amounts it will cause the animal to escape. Such a stimulus is unlikely to occur in two mice that are living together and well adjusted to each other. Pain is much more likely to occur when two strange animals meet and highly stimulate each other. There is no evidence at any age of an internal stimulus to fight; but agonistic responses to external stimuli change as the animal grows older, indicating that internal physiological conditions may have changed.

Hormones and fighting. One such change is the appearance of the male sex hormone. In house mice of the ordinary domestic strains, the females rarely if ever fight, but combats between strange males are frequent and can easily be induced in early maturity. Elizabeth Beeman tried the effect of removing the male hormone by castration at various ages. After they had rested twenty-five days, she put operated males together, but none of these animals ever fought. She then implanted testosterone propionate pellets in each animal to replace the male hormone. After this they fought like normal males. When the pellets were removed, most of them immediately stopped fighting, but a few kept right on. This showed that the male hormone did not control fighting completely.

She next tried the experiment of allowing animals to become accustomed to fighting each other in a series of round-robin matches. After this they were castrated and tested for fighting without any rest. These animals continued to fight without interruption. The male hormone must be present in order to get the animal to start fighting, apparently acting by lowering the threshold of responsiveness to the painful stimulation which normally causes fighting, but is not necessary once a strong habit has been established. Although small amounts of male hormones continue to be secreted by the adrenal glands of even castrated males, habit formation obviously has a strong effect.

The male hormone has no effect upon the fighting of adult females, as King and Tollman found when they injected it into castrated adults. The females responded with increased sexual behavior but no fighting. Later Bronson and other workers found that if they treated newborn female mice with male hormones, and then repeated the dose when they became adults, many of them fought just like males. Thus, there is a critical period in which the nervous system can be sensitized to the effects of the male hormone.

In normal development the gonad-stimulating hormones of the mother reach the fetus before it is born and stimulate the gonads to activity, with the result that a male mouse is born producing androgen. This acts upon his nervous system in the few days that the effect persists, and results in the fact that the male and female nervous systems become essentially different in adult mice. Still another hormone, cortisone, may be important in agonistic behavior. This evidence comes from studies on the causes of surgical shock. When a rat is severely injured—and probably a slight injury produces the same effects in a less obvious way—it first goes into a phase of shock in which body efficiency is lowered by a rapid and fluttering heart beat, decreases in body temperature and muscular tone, a decrease in blood sugar, and so on.

These symptoms may be reversed and body efficiency increased in the phase of countershock which may follow within a few minutes. Efficiency is then maintained until the animal recovers from the injury or becomes exhausted and dies. These reactions are controlled by hormones, the primary one of which comes from the pituitary gland. This stimulates the adrenal cortex to produce cortisone, which produces the physiological effects described above. The same reactions should occur in fighting whenever there is pain or injury, and it is possible that the tendency toward death-feigning is in part the result of shock. The direct study of the relationship between cortisone and fighting has not yet been made. Would cortisone make mice into better fighters or help beaten animals to resist their conquerors?

The original hormonal reaction in the pituitary is apparently set off by fluids which are released by injured tissues. Here we have a case where fighting might be influenced by internal metabolic reactions, but these changes are themselves produced by injury which must come from the outside. We must conclude that normal metabolic activities and changes have little if any influence on agonistic behavior and that it is different from ingestive behavior in this respect.

The effect of experience upon physiology.—Male mice reared and kept together in small groups may live peacefully together for months. Even if they are exposed to strange mice that have been reared under similar conditions they do not always fight. However, if mice are reared alone, they become progressively more irritable and more likely to fight if brought together. Along with this there are progressive changes in neuroamines, the proteins found in the brain. The mice appear to be emotionally disturbed by isolation and this in turn brings about changes in physiology.

Such changes are even more striking in the case of mice that have actually fought, and particularly so for the defeated member of a pair. Brain chemistry is modified, and the level of steroids in the circulating blood is profoundly changed, sometimes for periods as long as 24 hours. Similar changes can be brought about in a previously defeated mouse by the mere sight of its attacker. Thus the previous experience of the mouse brings about internal changes. Whether or not these hormonal changes act as stimuli to modify future behavior is unknown, but they presumably have some effect. Experience, of course, modifies behavior in other ways through the process of learning, but this will be discussed in another chapter.

Nervous control of fighting. Mice, being small animals, are ideal subjects for the study of fighting. But their very smallness makes them poor material for studies of the nervous system, and much of our detailed information has been derived from experiments on cats.

Cats are characteristically aggressive animals. Most of the play of young kittens consists of mock fighting. Unlike mice, male cats get into serious fights over the possession of females. Even in mating, the male and female will yowl, spit, and claw at each other in a way which is difficult to distinguish from conflict. Whenever a cat is threatened by a dog, we see its characteristic pattern of defense, arching the back and holding the tail erect. The hair stands stiffly on end all over the body, and the tail resembles a bottle brush. The cat growls, spits, and extends its claws, ready to strike if the dog comes near. If actually attacked, it becomes the image of fury, biting and clawing with all four feet with incredible rapidity. Such behavior is easy to stimulate in a cat by any sort of rough handling, as veterinarians find to their cost when trying to give pills or injections.

One of the important clues to the internal causes of this behavior was discovered by Bard when he removed the cerebral cortex of the cat brain by surgery. After his experimental animals recovered from the operation, they could live fairly normally except that they had to be fed by hand, presumably because the part of the brain which had to do with learning to eat was gone. When they were picked up or touched, they would immediately act like angry cats, spitting and scratching in all directions. The cats were hyperexcitable for fighting behavior, and a slight touch would set them going. The only difference between their behavior and that of normal animals was that their rage was poorly directed.

Another way of analyzing the function of the brain is direct stimulation with an electric current. In order to reach the inner regions, Ranson inserted very fine platinum electrodes in the brain of anesthetized cats. After an animal recovered consciousness, the experimenters ran a weak electric current through each electrode in turn. Some of the electrodes produced all the reactions of an angry cat. It turned out that these were always the ones that had been inserted in the hypothalamus, a region deep in the lower part of the forebrain.

Both the cerebral cortex and the hypothalamus are concerned with the expression of anger. Removing the cerebral cortex piece by piece shows that part of it represses anger, so that slight stimulation does not upset an animal in ordinary life. Another part actually excites anger. Normally the two parts keep in balance, which outside stimulation can tip in either direction.

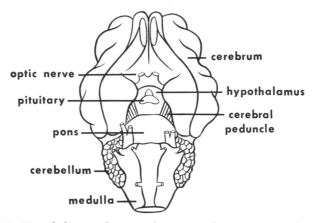

Fig. 11. Hypothalamus of a cat, showing its location on the lower side of the brain. This small part of the nervous system, so important in the physiology of the emotions, is located on the floor of the second part of the brain (the diencephalon), closely connected with the pituitary body.

Note also the position of the much larger forebrain or cerebrum. Surgical removal of its outer part or cortex causes "sham rage." (Modified from Ranson)

The hypothalamus itself is a brain center which acts to magnify and prolong the effects of the primary external stimuli which produce fighting behavior. This action is normally kept in check and directed by the cerebral cortex. Stimulation does not originate in the hypothalamus, since even when the cerebral cortex is removed, the original stimulus has to come from the outside, and the effect eventually dies out. From the human viewpoint, the hypothalamus is probably the place where the *sensation* of anger originates.

The hypothalamus is also the anatomical center of a number of adjacent structures in the brain, such as the amygdala, hippocampus, and septal area. Together they are known as the limbic system, whose function involves various sorts of emotional responses. Interestingly, the limbic system has changed very little in the evolution of the vertebrate brain, indicating that the fundamental emotional responses were developed very early in evolutionary history.

From the hypothalamus some stimulation goes to centers controlling voluntary muscles involved in scratching and body posture. A larger share goes to the sympathetic nervous system, which in turn activates several internal organs. The heart beats strongly and rapidly, digestion stops, and blood under high pressure is directed toward the skeletal muscles. As Cannon pointed out, these responses put the animal in readiness to deal with an emergency requiring great physical activity. Along with everything else, the adrenal gland is stimulated to release adrenalin, a hormone which affects the internal organs in the same way as the sympathetic nerves, but whose action is later and more lasting.

This is a very brief physiological description of the emotion of anger. As all of us know from personal experience, such activities in our own bodies are accompanied by certain sensations which themselves may stimulate fighting behavior. The question now arises: Do these sensations come from the activities of internal organs like the heart, or do they come from the central nervous system itself? One answer is given by the fact that adrenalin is often given to human patients for the treatment of hay fever and allergies. These patients report that they do not feel angry. Indeed, they feel very little except a slight sense of well-being and the fact that their nasal passages and lungs become clear. The sensations of anger do not follow the injection of adrenalin. Stanley Schachter has demonstrated the same point experimentally. He told human subjects that they were being given a vitamin preparation and then paired them with other subjects (actually experimenters, although this was unknown to them) who either insulted them verbally or put on a comedy act and tried to involve the subjects in it. They afterward reported that their emotions were either those of anger or elation, depending on the external situation. Actually all the subjects had received doses of adrenalin, which obviously had no consistent effect on their emotional states. Also, Walter B. Cannon long ago reported that patients who had their sympathetic nerves cut could still feel angry. It is apparent that whatever the sensations of anger may be, those which come from internal body activities are relatively unimportant and that the principal sensation comes from somewhere in the central nervous system, probably the hypothalamus and surrounding centers. However, the original stimulus must come from the outside.

In fear, another kind of emotion associated with agonistic behavior, the sensation of internal organs in activity is more apparent. Fear is usually associated with escape behavior rather than with fighting, but physiologists were long unable to distinguish it from anger by objective measurements. As anyone can tell you, the two emotions feel subjectively quite different, particularly in the intestinal tract. Persons who are afraid feel butterflies in the stomach and an urge to evacuate the bowels, whereas these symptoms never appear in real anger. There is some evidence that two forms of adrenalin with slightly different effects are secreted in fear and in anger. In any case the two kinds of emotion are closely related, as are the appropriate external behavior patterns of escape and fighting. In the fearful kind of emotional reactions, the sensations of the stimulated internal organs are distinctly recognizable and could have considerable effect as internal stimuli of behavior. Not as much has been done with the brain centers controlling fear as with those involved in anger, but removal of part of the cerebral cortex known as the gyrus cinguli will cause monkeys to become completely docile and show no escape behavior.

It may be concluded that agonistic behavior, like ingestive behavior, has a complicated network of internal causes. However, these causes appear to be quite different in nature. Hormones are much more important, and changes which are due to normal metabolic activities seem to have almost no effect. The chain of causes which lies behind fighting and escape behavior seems always to begin with some external environmental stimulus. There is no evidence for any true "hunger" for fighting.

Physiological counteraction and interaction. Cannon pointed out that many of the physiological mechanisms of hunger, fear, and rage are incompatible because of the antagonistic effects of the different parts of the autonomic nervous system in mammals. For example, the stimulation of the sympathetic nerves in anger tends to inhibit the contractions of the stomach and hence to

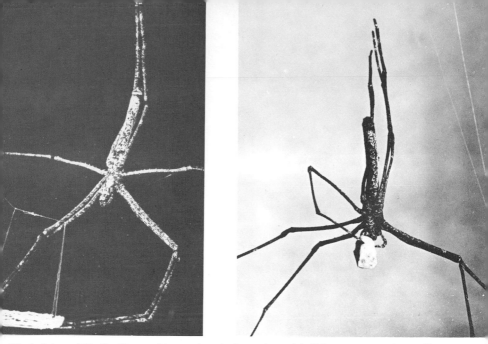

Manipulatory skill of spiders and insects in spinning webs and building nests exceeds anything known in vertebrates. Purse spider is shown *at left* with simple silken trap held between two pairs of legs, and *at right* with trapped fly that has been rolled up in silken thread. Spider must spin new web for each victim. (Photographed in Panama by R. Buchsbaum)

imal constructions. Paper nest of the white-
ed hornet (*Vespa maculata*) hanging from
 branch of a pine tree. The nest contains the
 od combs for the larvae; this particular one
 approximately 30 cm. in length. (Photo by
 P. Scott)

A smaller hornet's nest with part of the outer covering dissected away after the inhabitants had left. In the center are a number of cells in which the larvae are raised, all hanging down from the central stem. These are protected from the weather by several concentric layers of paper. All of these structures are built by finely coordinated movements which make the efforts of the larger vertebrates look crude by comparison. (Photo by J. P. Scott)

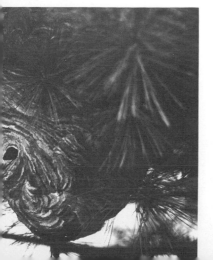

Animal constructions. Old beaver dam and pond. Dam is about 2 meters high and at least 50 meters long. Such dams are among the largest and most complex structures built by nonhuman mammals. The beavers cut trees along the bank, eat the bark, and use the logs to build the dam.

Chimpanzee nest. This is small (about 1 meter in diameter) and crude in comparison with the work of the beaver. Such nests are temporary sleeping places and are built each night. (Photo by Henry Nissen)

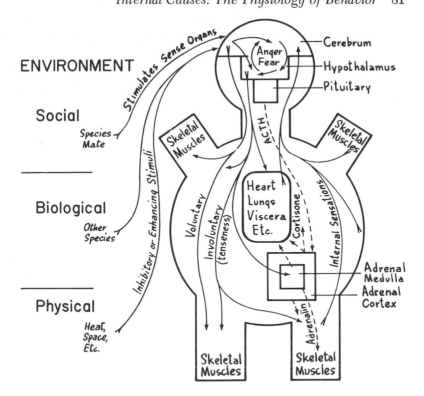

Fig. 12. Physiological mechanisms affecting agonistic behavior (social fighting) in a mammal. Hormonal effects shown in dotted lines, neural stimulation in solid ones. External stimuli on left excite sense organs; these stimuli are carried to the cerebrum to become sensations, and to the hypothalamus to become magnified as emotional sensations of anger and fear. Involuntary neural and hormonal stimuli are then carried to various internal organs, which may give some feedback to the cerebrum. Unlike the mechanisms underlying ingestive behavior, changes arising from cell metabolism have no major stimulating effect. (Original)

prevent the sensation of hunger. Fear interferes with hunger in another way, by stimulating violent activity in the stomach, which may give a sensation of nausea. On the other hand, many of the physiological reactions of fear and anger can be combined.

Other more complicated types of interaction may occur. If we give two well-fed mice a pellet of food, they will pay no attention to it and will not molest each other. The same two mice if hungry will struggle violently for a bit of food. In this case hunger has become a secondary cause of fighting, albeit a kind of fighting directed toward the possession of food rather than injury of the other individual. While fighting is primarily produced by external stimulation, it can sometimes be affected by changes which originate internally, as in this experiment.

There are many other ways in which networks of causes may interact, both through direct physiological means or through learning. This means that different kinds of internal stimulation can sometimes be added together, resulting in an unusually high state of excitement. One way in which this can occur is through the "ascending reticular activating system," a network of fibers in the form of a cord connecting the ventral part of the brain stem with the thalamus. Any kind of external stimulation activates these fibers, and the most general result is a state of arousal, the opposite of sleep, that can be measured by changes in the electroencephalogram as well as be seen in the behavior of the animal.

SEXUAL BEHAVIOR

The internal causes of sexual behavior are even more strongly associated with physical maturity than are those of agonistic behavior. Recognizable sex behavior in female sheep and goats first occurs at the time of the first estrus, usually after the animals are fully grown. The onset is more gradual in males, but the tendency to mount other animals is greatly increased in adults. There must be some kind of internal changes associated with sexual maturity, and what these may be is well illustrated by experiments with the sex behavior and the endocrine glands of the guinea pig.

Mating behavior of the female guinea pig. The sexual behavior of the guinea pig has been so thoroughly studied by W. C. Young that the details of stimulation are better known for this rodent than for almost any other animal. The female guinea pig first begins to show signs of mating behavior at about forty-five days of age. One of the first signs is increased restlessness and greater activity. Then for about eight hours the animal shows true sexual behavior. At first the female acts very much like a male, mounting other females and males if they are present. This sort of behavior is very exciting to the male, who tries to mount and is eventually accepted by the female. She then assumes a posture which will permit copulation. The female remains receptive for about eight hours and then abruptly stops. Thereafter she kicks and runs away from any male that approaches her and will not assume the sexual posture. If fertilization has not taken place, the cycle will be repeated in approximately sixteen days.

It is difficult to find any internal changes which occur quickly enough to act as nervous stimuli. There are changes in the cellular structure of the vagina and uterus before the behavior appears, but these occur very slowly; and it is difficult to see how they could result in nervous stimulation. Cellular changes are, in turn, caused by an increased amount of the hormone estrogen. If this hormone is injected into a non-receptive female, sexual behavior will follow. It seems to act by making the animal much more sensitive to external stimulation of any kind. Even when completely alone, a female apparently becomes hypersensitive to whatever stimuli are present in the environment, and her activity increases. The hormone seems to lower the threshold of stimulation, so that when a male appears sexual behavior is very easily elicited. Once mating begins, there is of course an intense reciprocal stimulation between male and female.

A peculiarity of the sexual cycle of the female is that it almost always occurs in the middle of the night, reaching its height in the very early morning. This made the guinea pig very inconvenient for the study of sex behavior until Professor Young found out that if he kept the animals in a dark room and turned on an artificial light at night, the cycle became reversed after a few weeks and the behavior could be studied at a more convenient time. It appears that changes in illumination act in some way upon the pituitary gland to produce a hormone stimulating the ovaries, which in turn secrete the hormone estrogen at the proper time. Similar effects of changes in light conditions have been shown to account for seasonal occurrence of sexual behavior in many birds and mammals.

Male sexual behavior. The sexual behavior of the male guinea pig is a much more constant affair. Any time a new female is introduced into his pen, the male will react to her, nosing her genitalia and making brief attempts at further sexual behavior. If the female is in estrus, the male becomes highly excited and walks around the female, throwing his haunches from side to side and making a sort of rumbling noise. The testes are extruded into a more prominent position, and he finally mounts and makes rapid pelvic thrusts. Ejaculation follows, and the male retires to groom himself. For an hour or so afterward, he reacts indifferently to this or any other female in heat.

One peculiarity of the guinea pig is that the male has unusually large accessory sex glands. When the fluid from these is ejaculated, it coagulates to form a hard copulatory plug which effectively closes up the vagina of the female for a considerable period afterward. This may serve to insure the survival of sperm, but it also tends to prevent any further effective sex behavior in the female. A copulatory plug is found in many rodents but is perhaps best developed in the guinea pig. It is possible that both the satiation of the male and the vaginal plug are mechanisms which prevent prolonged sexual behavior, which would of course be a dangerous luxury for any small and defenseless animal like a rodent.

When we look for any internal changes which might affect the sexual behavior of the male guinea pig, the only obvious possibility is pressure resulting from the accumulation of fluids in the accessory sex glands. However, when a castrated male is injected with male hormone, it begins to give the mounting response a day or so before enough fluid accumulates in the sex glands so that ejaculation is possible. Just as in the female, the chief internal factor affecting male sexual behavior is hormonal, and it seems to act by lowering the threshold of stimulation rather than by exciting a nervous stimulus itself.

There must also be some sort of physiological satiation mechanism in the male to account for the fact that once ejaculation is complete, the male can no longer be stimulated by a receptive female. Fatigue may enter into the picture, or it may simply be that the activity of the sex organs produces inhibitory stimuli. It also requires a certain amount of time before the fluids in the accessory sex glands can be secreted again. This period of satiation is not typical of all mammals, as bulls will repeatedly mount receptive cows for hours at a time, even after the seminal fluids have been completely exhausted, provided a series of fresh females is introduced or novelty is brought into the situation in some way. This is another illustration of the principle that a stimulus is a change.

In an adult male, the male hormone is constantly circulating in the blood. While it provides a necessary condition for sexual behavior, it is difficult to see how it could act as a stimulus, as this would violate the principle that a stimulus is a change. On the other hand it is possible that the male hormone might enter into a chemical combination with certain cells in the brain and thus continue to stimulate them. A. E. Fisher tested this hypothesis by injecting the male hormone into various areas of the brains of male rats. The results were surprising. In some cases he did indeed induce malelike sex behavior which persisted over long periods, but rats also reacted by exhibiting maternal behavior, attempting to build nests and carry young about. In some cases a male rat attempted to carry adult females as it might young infants. The results, while they indicate that the male sex hormone has strong stimulating properties, also lead to the conclusion that the male hormone is not a specific stimulus for male behavior but rather has a general effect of lowering the threshold of stimulation for several kinds of behavior.

As with agonistic behavior, the hypothalamus probably functions to magnify and prolong the effects of stimuli leading to sexual behavior. Continuous electrical stimulation in a male rat can maintain sexual responses for as long as an hour. Conversely, the hypothalamus becomes electrically active during sexual behavior, as measured by the electroencephalograph. If we postulate some sort of feedback mechanism, the hypothalamus could function to prolong sexual stimulation.

We can see that behind sex behavior there is still another set of physiological causes than those which affect ingestive and agonistic behavior. In contrast to agonistic behavior, hormones applied to the hypothalamus stimulate sexual behavior. Very small doses of estrogen injected into this part of the brain will cause a female cat to show the behavior of estrus; that is, increased activity and receptivity to males. Internal changes that are not the direct effect of external stimulation thus affect sexual behavior.

CARE-GIVING BEHAVIOR

Parental care of the young is widespread in highly organized animals and in some ways reaches its greatest complexity in birds. These animals must protect and feed the young from the egg stage on, whereas in mammals such early care is purely physiological and takes place in the uterus before the young are born.

Parental care in the wren. One of the favorite birds of house-holders is the wren. These tiny birds have an attractive bubbling song, and since they nest in holes, they can be readily attracted to our back yards by setting up birdhouses. The entrance must be of the right size, or the house will be taken by larger birds. As with red-winged blackbirds, the males come back each spring and set up territories before the females arrive. If we put several nest boxes close together in the hope of attracting whole tribes of wrens, the male comes around to each box in his territory, busily throwing out the remnants of former nests and fighting off any males which try to move in.

Then the male starts bringing in coarse twigs to one of the boxes and begins to rebuild the nest. When the female arrives, she takes over the job and completes the nest lining. After this she begins to lay eggs, one each day until an average number of six have been produced. As the eggs are laid, she begins to sit upon them, longer and longer each day. When the clutch is complete, she starts a period of constant incubation which lasts until the eggs hatch twelve or fifteen days later. If the female leaves the nest for a short time, the male usually stays close by as if on guard, although he does not sit on the eggs himself. When she returns, he goes off to feed.

After the eggs are hatched, both parents start feeding the young and may make as many as two or three hundred trips per day, bringing insects back in their bills and stuffing them into the mouths of the baby birds. Meanwhile the young birds still need to be protected, and the female spends part of her time brooding them. Sometimes the female will desert the nest. When this happens, the male may take over the entire care of the young.

Professor Kendeigh of the University of Illinois has studied this behavior in great detail by means of an instrument called the itograph. A perch arranged like a seesaw is placed at the entrance to the nest. Each time a bird goes in or out it presses down a small switch which is connected with a recording instrument. In this way the number and direction of visits to the nest can be measured very accurately. While the eggs are being incubated, the female visits the nest less frequently if the temperature is very high. Also, if the temperature falls very low, the female uses up so much heat that she has to spend more time away from the nest looking for food. It is obvious that environmental changes have some effect on this behavior, but there must be internal factors as well, to account for its sudden appearance. As in the broody hen, epimeletic behavior begins long before the eggs or young appear. Going along with the changes in behavior are seasonal changes in the reproductive organs; and it is probable that hormones have a great deal to do with parental behavior. The hormone prolactin appears in the female just as she stops egg-laying and begins to brood.

This hormone has certain very special effects in pigeons and doves. Like other members of the family, male and female ring-doves are very much alike in both appearance and behavior. After pairing, both parents cooperate to build a nest. After the eggs appear the male sits on the nest for six hours during the middle of the day while the female goes off to feed. When the young are hatched both parents feed them by regurgitating "crop milk," a substance secreted by the lining of the crop. In an ingenious set of experiments, Daniel Lehrman and his colleagues at the Rutgers University Institute of Animal Behavior analyzed the effects of hormones on this behavior. Ordinarily the birds will not sit on the eggs until approximately a week after courtship starts, even if eggs and a nest from other doves are placed in their cage. Injecting the hormone progesterone into either males or females induces immediate sitting on the nest. The hormone estrogen is less effective, producing incubation behavior one to three days later. Finally, the hormone prolactin, while ineffective in producing incubation, greatly increases the parental behavior of regurgitating crop milk for the young, although it is not necessary for the behavior of regurgitating, which can take place even though milk is not present. This is the same hormone that precedes milk secretion in mammals, and in doves has the physiological effect of activating the lining of the crop so that crop milk can be produced. A female ringdove can be stimulated to lay eggs by the sight of a male separated from her by a glass plate. Likewise the crop of the separated male will develop the capacity to produce milk if he can see his mate sitting on the eggs, but not if she is concealed from him.

The whole process of reproduction and parental care in these birds consists of an elaborate interaction between behavior and hormones. Courtship behavior stimulates the pituitary glands of both sexes which then produce the gonad-stimulating hormones which trigger the production of eggs and hormones from the sex glands, which in turn produce incubation behavior. The net result is that the behavior and physiology of the two parent birds become coordinated in a highly adaptive fashion.

Enough has been said to indicate that while the same hormones influence the reproductive behavior in different vertebrate animals, their physiological functions may vary widely, and the organization of behavior with respect to hormones may be quite different in different species.

Maternal care in mammals. In mammals the secretion of milk produces a different physiological situation from that in most birds and even from members of the dove family. A mother dog whose pups have been taken away becomes very uneasy and restless as the mammary glands become filled with milk, and this excitement can apparently be considerably reduced by applying irritating substances which inhibit the flow of milk. In the case of milk secretion, we have a definite physiological change which can act as a nervous stimulus.

This is not the only factor. Many female rats will show maternal behavior even when not pregnant and not lactating. The response of picking up the young and carrying them around may be shown even by males. Newborn rats seem to act as a primary stimulus for this kind of behavior, and the strength of the reaction can be greatly increased by the injection of the hormones prolactin and progesterone. Prolactin eventually induces the flow of milk, which itself may have some effect on the behavior. It would appear that hormones have important modifying effects on maternal behavior in mammals, but the effects of different combinations and dosages are so complex that no clear picture has been obtained. In the rat, maternal behavior can be induced in virgin females by injecting blood plasma from females that have recently borne young, but the nature of the hormones carried in this blood has not been determined.

The activity of hormones in sexual and care-giving behavior is quite different from that in fighting. The secretion of adrenalin actually stimulates or inhibits the activity of internal organs like the heart and stomach. For the sex hormones and prolactin there is no evidence that any internal muscular activity is stimulated, but rather that internal glandular activities are increased and that individuals become more susceptible to outside stimulation. How then can we account for the fact that an isolated female rat will become more active during the period of normal estrus or when hormones are injected? Some sort of internal change or stimulus must account for this behavior. It is possible that the animal simply becomes more susceptible to any kind of outside stimulation, however slight. It is also possible that the hormones directly stimulate brain centers which control stimulation for sexual behavior, but this is difficult to reconcile with the general principle that stimulation must consist of a change.

OTHER SYSTEMS OF BEHAVIOR

Our knowledge of the physiological mechanisms lying behind behavior is scattered and incomplete. Some systems have been very thoroughly studied in a large variety of animals, but others are still almost unknown. In shelter-seeking behavior, the physiological changes which take place in reaction to cold are well known in mammals. The mechanisms of shivering and redistribution of the flow of blood have an obvious relationship to behavior. Likewise, the physiological changes in the bladder and bowels which act as stimuli for eliminative behavior are well understood.

There is still much to be learned regarding the physiology of investigative and allelomimetic behavior. Novelty plays an almost overwhelming role in the elicitation of investigative behavior, and there is little evidence that internal stimulation plays a strong part.

In the case of allelomimetic behavior, we have a clue to an emotional mechanism which accounts for the development of motivation toward the behavior of doing what other animals do. A young dog, beginning at about three weeks of age, will exhibit distress vocalization if separated from its litter-mate companions. The reaction begins almost immediately and is almost as quickly relieved by restoring the familiar companions. In this way the puppy must quickly learn that it can avoid the uncomfortable emotional response by remaining with other animals and that in order to do this he must do what they do. There may also be a positive pleasure associated with allelomimetic activities, but this has not yet been demonstrated. The physiological changes which accompany this emotion are still unknown, as are the brain mechanisms which control it.

The evidence which we now have shows that internal changes and causes of behavior differ with each major behavioral system, sometimes in detail and sometimes in entirety. This means that an understanding of the internal causes of behavior must be based on the direct study of the particular behavior patterns involved and that it is unsafe to make generalizations from one system to another. What is called "emotion" or "drive" in one kind of behavior may be physiologically completely different from that found in another.

In the same way, causes of behavior may differ from species to species. Examples of the physiological causes of behavior have been drawn chiefly from the higher mammals and certain birds because it is on them that the most detailed studies have been made. While some of these physiological mechanisms may be fairly widespread, there is no ground for assuming that they are universal in the animal kingdom. On the contrary, the mechanisms underlying sexual behavior in an insect, where air-borne pheromenes may affect behavior in ways that are similar to the effects of blood-borne hormones, are very different from those of a vertebrate. Although we have evidence that some kind of internal stimulation is important in every major group of animals which show active behavior, the behavior of a species can be thoroughly understood only when its special physiology has also been studied.

SUMMARY

This chapter is concerned with the internal causes of behavior which are frequently lumped together under the terms of "motivation" or "drive." If we study these causes in detail, we find that there is a complex network of internal causation peculiar to each general system of behavior. In ingestive behavior of mammals, the sensation of hunger comes from the violent contractions of an empty stomach. These in turn are influenced by a low level of glucose in the blood, which may also produce an unpleasant sensation in other parts of the body. Glucose also affects a hunger-regulating center in the brain, since operations on the hypothalamus of a rat can produce overeating. There are other special physiological mechanisms which affect the choice of diet with regard to salts, vitamins, and water. The outstanding feature of the physiology of ingestive behavior is the fact that changes occur within the body as a result of using up food substances, and that these internal changes stimulate behavior. In other words, part of the stimulation for eating has an internal origin.

Fighting, on the other hand, does not seem to occur in the absence of external stimulation. In the mouse the primary external stimulus for agonistic behavior appears to be pain. A very young mouse reacts by trying to escape. As it grows older the reactions change to defensive and offensive fighting, correlated with the development of teeth and sense organs and the appearance of the male hormones.

The presence of the male hormone in an adult mouse lowers the threshold of stimulation with regard to fighting but does not seem to affect activity in the internal organs. On the other hand, both adrenalin and cortisone are produced during fighting behavior and directly stimulate internal organs like the heart and stomach.

The internal nervous control of fighting has been better studied in the cat than in other animals. Experiments with brain surgery lead us to the conclusion that there is a mechanism involving both the hypothalamus and cerebral cortex, which acts to magnify and prolong the effects of external stimulation. Cats which have part of the mechanism missing are either hyperexcitable or overplacid. In fighting, all the evidence indicates that the original stimulation must come from the outside. There are complex internal mechanisms for transmitting, modifying, and prolonging this stimulation, but all of them are dependent upon external causes.

In sex behavior the internal mechanisms are different from either of the other two. Internal changes in hormonal levels occur at the time of sexual maturity in both sexes in mammals, and cyclical fluctuations occur in females. The sex hormones chiefly act by lowering the threshold of external stimulation, but they also produce increased restlessness and activity through some sort of direct stimulating effect upon the central nervous system.

The effects of hormones in care-giving behavior are very similar to those in sex behavior, except that one result in mammals is the secretion of milk. This produces an important internal sensation which stimulates maternal care in somewhat the same way as stomach contractions stimulate eating.

All this information has great significance with regard to the problem of the control of behavior in man and animals. A very important part of the causes of ingestive behavior lies in a special kind of internal stimulation which is the result of living itself and hence not under direct control. We cannot assume that this is also true of other kinds of behavior. Agonistic behavior is apparently entirely controlled by external stimulation, and it should be possible for either men or animals to live peaceably for long periods in the proper kind of environment. There is no evidence for a hunger to fight.

The situation is somewhat intermediate in sexual behavior. The sex hormones make the individual extremely sensitive to external stimuli. Even contact with inanimate objects may be stimulating. As a result even inexperienced animals become highly stimulated to attempt sexual behavior, and motivation is afterward strengthened by habit formation. It is possible that there is some internal stimulation of the central nervous system by sex hormones, but compared to ingestive behavior, external stimulation is much more important. Care-giving behavior is similar in that special hormones also lower the thresholds for external stimulation.

The evidence indicates that we can arrange different kinds of behavior in a crude spectrum with regard to the importance of internal stimulation, with ingestive behavior at one end and agonistic at the other, and sexual and care-giving behavior somewhere between. The ease with which behavior can be subjected to external control varies on the same scale.

We conclude that the mechanisms of internal stimulation are highly complex. The concepts of motivation and drive, while useful in general description, are too simple to explain the actual facts. The idea that an animal has so much "drive," which can be measured like the pressure of steam in the boiler of an engine, is too simple to be useful. Instead we find that internal stimulation varies with the kind of behavior involved, with the physiological mechanisms peculiar to the animal, and with the amount of external stimulation in the surrounding environment. This last variable introduces another, the effect of previous stimulation. As we shall see in the next chapter, its effects tend to be both additive and permanent, making it one of the most powerful sources of internal excitation.

CHAPTER 5

Learning: The Effects
of Experience

The complicated nature of experiments on behavior has sometimes led to confusion and occasionally to humor. There is an old
joke among biologists known as the "Harvard law" of animal
behavior: "When stimulation is repeatedly applied under conditions in which environmental factors are precisely controlled the
animal will react exactly as it pleases." This somewhat ponderous
epigram chiefly reflects the common frustration of experimenters
who find that animal behavior does not conform to their pre-
conceived ideas. But it is also related to a fundamental principle.
If it were repeated with the conclusion that "the animal reacts
differently each time," it would accurately express the basic
principle of variability of behavior.

VARIABILITY AND LEARNING

Variability may be seen even in primitive animals. Anyone who
has watched paramecia under the microscope has seen them run
into obstacles, back off, and start off on a different path. They
never back off and ram into the obstacle in exactly the same
fashion. If they did so, they would never succeed in adjusting to
the situation. It is obvious that a necessary part of the process
of adaptation in the avoidance reaction of the paramecium is
variability.

Some of the variability of behavior in animals is due to variability in kinds of stimulation, but a good deal of it is an essential
part of adjustment. Behavior is never a rigid mechanical affair;
even such a simple element of behavior as a reflex shows a surprising amount of variation. Any species has a variety of responses
which may be tried as alternate possibilities. Even a paramecium,

which has a limited repertory of behavior, may sometimes squeeze under a strand of algae instead of bouncing off to try another direction.

We can conclude that there may be variability both in a particular response and in the general type of response which is given to any stimulus. When a paramecium collides with an obstacle, it may either give the avoidance response, which is itself variable, or display an entirely different kind of behavior. Such variability is typical of even the simplest animals. With their greater mobility and more complex motor abilities, higher animals usually show a much larger variety of responses, each one capable of considerable modification.

Variability is such a fundamental part of adaptation that the task of finding some regular laws which predict behavior seems almost impossible. There are, however, certain factors which tend to reduce variability, and one of the most important of these is learning or experience.

Learning is a popular term which has been used in a great variety of meanings both in literature and in science. It is used here as a matter of convenience to describe a widespread phenomenon in animal behavior: that when an animal is repeatedly placed in the same situation and stimulated, its later behavior is affected by what has happened before.

We can find such effects even in the lower animals. Among the numerous experiments described by Jennings is one in which he repeatedly dripped carmine particles on top of a stentor. This is a large funnel-shaped protozoan which, although it belongs to the same class as a paramecium, usually remains attached to some solid object, holding the large end of the funnel upward and drawing in a stream of water from which it strains out food particles. When carmine particles come instead, it turns to one side. If the experimenter changes the direction of the stream of carmine, the stentor again changes its position but cannot escape the particles. It then reverses the direction of its cilia and blows the carmine away for an instant. When this is repeated several times without success, the stentor finally contracts and retires into the protective tube around its base for something like half a minute.

So far the animal has shown typical variability of response to a stimulus, but now comes the interesting part of the experiment. When the stentor comes out, will it go through its whole cycle

Fig. 13. Behavior is modified by experience even in a one-celled animal like a stentor. *Left,* the stentor starts to take in carmine particles dropped by the experimenter. *Center,* as the flood continues, it bends to one side and blows the particles away by reversing its cilia. *Right,* this being ineffective, it contracts into its protective tube. The stentor has now tried out several attempts at adaptation, but only one is successful. When the stimulus is again repeated, the stentor responds with the last response rather than starting again with the first, showing that its behavior has been modified by experience. (Redrawn from Jennings, *Behavior of the Lower Organisms,* 1906, by permission of Columbia University Press)

of variable adaptation, or will its behavior be modified by what happened before? Actually, the animal contracts again as soon as the carmine particles touch it, and it may be concluded that a stentor has a "memory" of at least half a minute.

Such effects as these do not appear to be the result of fatigue or other metabolic disturbances, as can be experimentally demonstrated with higher animals. Rats which have learned a maze perform better than inexperienced animals, even after a four-month interval with no practice, and the length of time in which the effect can be demonstrated is limited only by the life of the rat and the patience of the experimenter.

The usual effect of repeated experience in the same situation is the improvement of adjustment to the stimulation. The stentor avoids the carmine particles more successfully, and the rat runs faster through the maze. However, improvement is not an invariable rule, for sometimes all attempts at adaptation may be unsuccessful. The most general definition of learning is therefore *the modification of behavior by previous experience.* As such, learning is one of the major causal factors affecting behavior.

Widespread as this tendency is throughout the animal kingdom, the evidence which supports general theories and laws of learning rests on a remarkably narrow basis. Very few species and very few kinds of behavior have been thoroughly studied. Most of the experimental work has been concerned with the ingestive behavior of a few higher vertebrates, particularly the dog and rat. Some of the findings appear to be basic and general, but some of them may only reflect species peculiarities. As pointed out in an earlier chapter, experiments are most meaningful when based on a thorough knowledge of the behavior of the species used. Before we attempt the modification of behavior through learning, we need basic descriptive information about the behavior to be modified.

Ingestive behavior in the dog. An adult dog is primarily a carnivore, and his ingestive behavior and physiology are both well adapted to eating meat. As with all carnivores, whose food supply is not constantly available, a dog is also well adapted for going without food for long periods. A healthy adult dog can go at least a week without food or water without serious harm. When he finds food, he eats hastily and chews it little unless it contains bones which have to be reduced to fragments. He produces copious amounts of saliva which probably help to make the large chunks of food slide down easily. A dog also vomits readily, and this is part of the normal pattern of behavior for feeding the puppies at one stage in their development.

The newborn puppy obtains all its food by nursing. At about three weeks of age it may begin to taste various odorous substances. Along with this it mouths and chews all kinds of objects, including the bodies of other puppies and pieces of wood and pebbles. Stronger reactions are given to certain kinds of food, particularly raw meat and the vomit of the mother. On the first experience with either type of food, hungry puppies react immediately and eat without hesitation. It looks as if there were certain primary responses to food which are given without previous experience.

Among these primary responses is the involuntary flow of saliva. This is a reflex touched off by the presence of food in the mouth. It is affected by later experience, because older dogs will begin to salivate at the sight of food before the reflex can be set off by direct contact. It is this one part of the complex patterns of ingestive behavior of the dog which was so extensively studied by the Russian scientist Pavlov.

Pavlov was a physiologist who was interested in finding out something about the function of the nervous system and particularly the brain. As a result he did a series of experiments which have profoundly affected the development of scientific thought on the subject of learning. His findings can be reduced to certain generalizations or laws. Although his original evidence was based on one very special behavior pattern in one species of animal, there is much evidence that the laws have a more general application.

The Pavlovian experiment. The dog which is to be the subject of the experiment is first given a minor operation in which the duct to one of the salivary glands is led to the outside of the mouth instead of the inside. The saliva secreted through this duct during the experiment can be accurately measured either by volume or by the number of drops counted. After the dog has recovered from the operation, he is placed in an experimental room upon a stand, with his legs encircled by loops attached to an overhead bar. He can move to a certain extent but will lift himself off the ground if he goes too far. When the dog has gotten used to this situation, experimental training begins. This consists of presenting food to the dog in connection with various sorts of signals, which may be anything from bells and whistles to circles and triangles on a square of white paper. A good deal of the success of the experiment depends on the cooperation of the dog. Some animals will not become accustomed to restraint and cannot be trained, while others may be so calm that they simply go to sleep and pay no attention to what is going on.

SOME GENERAL PRINCIPLES OF LEARNING

The primary stimulus. As stated above, a very young puppy at the age of three or four weeks, who has had no previous experience with solid food, will give an immediate reaction to meat and try to eat it and even keep other puppies away. A similar instantaneous reaction is given to food freshly vomited by the mother. These kinds of foods may be thought of as *primary stim-*

uli which automatically produce eating behavior without previous experience. Pavlov found that showing a dog meat or allowing it to eat would cause it to drool and hence drip saliva from the duct which opened to the outside of the dog's mouth. He also found an instantaneous reaction to weak acid placed in the mouth, which is interesting in view of the fact that vomited food is acid from the stomach secretions.

This idea of a primary stimulus which produces a reaction without preliminary training is an essential part of the theory of learning. It is reasonable to suppose that even in animals capable of a great deal of learning there is some inborn tendency to react to certain kinds of stimuli. Otherwise there would be no behavior which learning could affect. It should not be supposed, however, that there is always just one kind of stimulation which will produce a certain response or that an animal will always give an invariable response to a primary stimulus. This would conflict with the principle of variability referred to earlier. Although an animal has a strong tendency to give a particular reaction to a particular primary stimulus, it may also give a variety of others. In Pavlovian terminology, the response to a primary stimulus is called an *unconditioned* (or *unconditional*) *response*. In the terminology of instinct theory, a primary stimulus is called a *releaser*.

The law of association. Having discovered a primary stimulus which regularly produced salivation, Pavlov tried the experiment of giving the animal a secondary stimulus which did not ordinarily produce salivation and then observing what happened under various circumstances. He sounded a buzzer just before he gave the dog some meat. The next time he sounded the buzzer, the dog began to salivate before it saw the food. After he repeated the experiment several times, this tendency became stronger, and the dog drooled copiously at the sound of the buzzer, even when no food was forthcoming. This illustrates the principle of association: that *a secondary stimulus which closely precedes a primary stimulus becomes associated with the responses normally produced by the primary stimulus.* This is what Pavlov meant by a "conditioned reflex"; that is, a reflex that had been affected by previous combinations of stimuli.

"Conditioned reflex" is an unfortunate term, since behavior is affected by *changes* in conditions rather than by conditions. However, the term is now so widely employed that it is necessary to understand it. Actually, it was mistranslated from the original Russian, where it read "conditional reflex," meaning a response that appears conditionally.

The effect of association is not ordinarily produced if the secondary stimulus comes *after* the primary one or if it precedes it by any great length of time. This phenomenon is closely similar to what human beings call the law of causation, which assumes that if two events always occur together and one precedes the other, the first is the cause of the second. The dog acts as if the buzzer were the "cause" of its getting the meat, as long as the two are closely connected and the buzzer comes first.

After Pavlov had produced an association between buzzer and food in a particular dog, he tried another secondary stimulus on the same animal. This time he rang a bell but gave the dog no food. The dog of course did not salivate. After this had happened several times, he sounded the buzzer and rang the bell at the same time. The dog produced some saliva, but much less than it usually did to the buzzer alone. It can be inferred that an association can be built up between a secondary stimulus and a lack of response. This may be called *the law of negative association or inhibition.*

Pavlov found that any secondary, or "neutral," stimulus that occurs repeatedly in a dog's environment can be shown to have an inhibitory effect. Apparently an animal can learn to do nothing just as well as it can learn to give a positive response to a stimulus. It does this simply by not reacting when stimulated. This means that training against undesirable responses can be accomplished without punishment, and has wide implications in training both child and dog. It is possible to housebreak a puppy with little if any punishment simply by making sure that it always urinates and defecates in the proper place. Simply by not urinating in the house it forms an inhibition against it. Likewise, in a child it is possible to accomplish negative training without punishment. In fact, this method of *passive inhibition* combined with positive stimulation of desirable behavior is the most powerful and effective means of controlling destructive aggressive behavior.

The law of generalization and discrimination. Pavlov found that associations could be set up with all sorts of secondary stimuli. If the stimulus was a whistle, the dog would afterward respond to all sorts of whistles, some with very different loudness and pitch from that which had accompanied the food. The dog generalized from one to a whole class of stimuli. If the experimenter repeated the various whistles but gave meat only when one particular kind of whistle was sounded, the animal very soon stopped responding to the whistles which were not followed by meat. The response, as Pavlov put it, was extinguished. The result was that the dog now discriminated between the different stimuli.

These principles may be stated generally as follows: If an association is formed between a particular secondary stimulus and a primary one, the response will tend to be given to any similar secondary stimulus. This is called *generalization.* If the primary stimulus does not in fact occur with such similar secondary stimuli, the generalized response will tend to die out or be extinguished, and the response will be given to only one secondary stimulus. This is called *discrimination.* We can easily see that a negative association or inhibition is built up with the other secondary stimuli and that discrimination can be explained as the modification of a generalization by the process of negative association.

Extinction and recovery. In still another kind of experiment, Pavlov first trained the dog with a combination of buzzer and meat and then started sounding the buzzer without giving any food. After a few trials the response dropped to zero. The next day a few drops of saliva might appear at the first sound of the buzzer. After this there was no more saliva, either then or on succeeding days. This is called the *extinction* of a response, and it can be explained as the modification of a positive association by the process of negative association.

However, when the dog was given a long rest from stimulation for several days or weeks, the response would appear again without any reward of food. This is called *recovery* from extinction. As far as this and other experimental evidence goes, it looks as if an association never completely disappears and that some very lasting effect has been produced in the nervous system. This effect can be masked by subsequent training, as in the process of extinction, but some traces of it appear to be permanent.

The strength of association. This may be tested by finding out how long it takes for a response to a secondary stimulus to disappear when no food is given. In general, Pavlov found that the duration of the response depended on the number of times that food was given in connection with the stimulus and also on the recency of the training. These results are quite similar to those which have been obtained in experiments with memory in human individuals, where it is found that the amount of material remembered depends a great deal on the frequency and recency with which it has been studied. The strength of an association therefore depends chiefly on the number of times in which the primary and secondary stimuli have occurred together.

The association of a stimulus with a response may also be called *habit formation*. This leads us to some interesting conclusions. If the principle of frequency is a general one, it would follow that it takes about as long to break a habit as it does to form one in the first place, and the principle of recovery indicates that a habit can never be wholly obliterated. An old dog can learn new tricks, but it is very difficult to make him forget his old ones, since these are habits formed by hundreds or even thousands of repeated associations.

Habituation. This is a somewhat different phenomenon from habit formation and reflects the fact that even a primary response will tend to become extinguished if it is not reinforced in some way. A young puppy of three or four weeks of age will suddenly lift his ears and draw back from any loud and sudden noise. If we experimentally repeat such a noise and no harm comes to the puppy, he will respond less and less violently and finally give no reaction at all. We would expect that the puppy would eventually form a strong association between such harmless loud noises and inactivity, although the startle response might be repeated if the stimulus was given after a long period of silence.

The physiological basis of habituation is unknown, but it has certain obvious resemblances to the phenomenon of accommodation of the sense organs, except that in the case of habituation the animal is obviously aware of the stimulus even though he does not respond to it. Habituation has the effect of causing primary responses to become organized with respect to situations that are meaningful to the animal. It also has the effect of greatly reducing the responsiveness of an older animal as compared with a young one. A young puppy rushes around the yard, quivering with excitement at any odor that reaches him on a casual breeze, whereas an old dog sleeps on a door step, occasionally opening one eye.

As we look at the general results of Pavlov's experiments, we can conclude that there seem to be only two basic principles, association and generalization, and one basic process involved: association of either the positive or negative type. Generalization, discrimination, extinction and recovery, and strength of association can all be understood as the result of this reasonably simple process applied to various combinations and permutations of environmental stimuli.

In all these results of Pavlov there is a suggestion that he was dealing with fundamental phenomena which have a very widespread application and occurrence. However, it is not scientific to assume that this is the case without examining a large body of evidence based on various kinds of behavior in many different species. Much of this evidence has been accumulated, but it still has a great many limitations.

THE EXTENSION OF PAVLOV'S RESULTS TO VOLUNTARY BEHAVIOR

One of the limitations of Pavlov's experiments is that he worked with only one example of simple reflex behavior. This raises the question whether the results also apply to voluntary behavior. Since the salivary reflex is a part of ingestive behavior, it would be logical to make further experiments on voluntary behavior associated with eating.

A reflex differs from voluntary behavior chiefly in the way in which it can be modified. The salivary reflex can be modified by learning so that small or large amounts of saliva are produced, and it may be set off or inhibited by various stimuli. A simple bit of voluntary behavior, such as closing the jaws, is no more variable, but it can be organized and related to behavior and stimulation in a much more flexible and highly adaptive way. A dog starts to salivate as soon as he sees the food, but he does not begin to snap his jaws until he actually gets it in his mouth. Back of this difference is a much more complicated set of nervous pathways than exist in a simple reflex.

We may now ask whether voluntary behavior is affected by the simple processes of positive and negative association, or whether its more complex nervous connections involve other processes of learning. This point was extensively tested by the American scientist B. F. Skinner, who devised a somewhat different kind of apparatus from that of Pavlov and who, in common with many American psychologists, experimented with the rat rather than the dog.

Ingestive behavior in the rat. In contrast to the dog, the rat's digestive system is primarily adapted for a herbivorous diet, although it will eat meat or almost any other sort of food if available. It has a very small gullet, so that the food has to be finely chewed before it can be swallowed. All this means that a rat, unlike a dog, has to spend a considerable amount of time eating, although it has been found in laboratory experiments that rats can keep themselves healthy by eating as little as an hour a day. Another difference from the dog is that it is difficult for a rat to vomit, a fact which is made use of in certain rat poisons like red squill, which other animals can easily eject from the stomach.

As part of their normal behavior, wild rats carry food around and deposit it in concealed places. However, there is no record that they ever deliberately feed these morsels to their young, which have to go out and find their own food once the nursing period is over. Rats get hungry after a few hours without food and show signs of starvation within a few days. They are therefore animals in which it is easy to arouse a strong internal stimulus of hunger, and this is one of the reasons why they have been made the subject of experiments by psychologists. It is very easy to get a rat to work for food.

In some Skinnerian experiments, the rat is first put on a restricted diet and its weight brought down to approximately 85 percent of normal. This means that it is almost constantly hungry. While in this condition, it is put into a box which has come to bear the name of its inventor. The box has nothing in it except a small bar which, when pressed, will release a tiny pellet of food into the cage. As the rat investigates its cage, it sooner or later steps on the bar and releases a small bit of food. Before long the rat is rapidly pressing the bar and eating the food. Since it never gets enough to satisfy its hunger, the rat will go on doing this for hours at a time. Meanwhile an automatic recording device graphs the total number of times that the rat has pressed the bar during any given period.

Using this technique Skinner got results very similar to those of Pavlov. The rat obviously forms an association when it learns to press the bar to get food. If the food supply is cut off, the rat soon stops pressing the bar, which is of course extinction or negative

Fig. 14. Skinner's method of testing the effects of learning on voluntary behavior. The rat presses a lever which releases small pellets of food. If the food is released only when the light is turned on, the rat soon learns to stop pulling the lever in the dark. Similar devices can be used for a variety of experiments on many species of animals. Elaborate electronic control and recording devices are shown following p. 112. (Original)

association. If food is given only when a light is turned on, the rat soon quits pressing the bar in the dark, having learned to discriminate between the two conditions. In this way Skinner has verified the supposition that the Pavlovian principles of association and generalization can be extended to kinds of behavior other than reflexes. In addition, his work shows that associations can be made between a response and *subsequent events* as well as with preceding stimuli. In other words, the rat learns that its behavior has results.

Learning and adaptation: The success principle. The chief difference between Pavlov's experiments and those of Skinner is that the salivary reflex is affected by what happens just before it, whereas the bar-pressing response of the rat is mainly affected by what happens afterward. This means that associations can be made between any two events that occur closely together in time, so that long chains of stimuli and responses can be built up. In Skinner's box the chain is built up in the following way: First there is an internal change or stimulus of hunger which leads to general activity and eventually to bar-pressing. This in turn causes a pellet of food to roll into the cage. The food then acts as a stimulus for the response of eating. This chain of events has finally led to a successful adaptation to the original stimulus of hunger.

We now see that there is a connection between the principles of learning and the principle of adaptation. A stimulus is a change, and a response is an attempt to adapt to a change. Responses tend to be variable, and the principle of association will tend to cause the animal to select a response which gives a successful adaptation. Put in the simplest terms, the processes of learning tend to perpetuate those responses which produce successful adaptation and eliminate those which are unsuccessful.

Motivation. Once an animal has begun to learn, the amount of motivation is directly affected by the amount of successful adaptation. Some motivation always comes from primary stimuli, such as food in the external environment and hunger inside the body. But this motivation is greatly magnified by training. An animal may work many times as hard for food after obtaining it on several occasions than he did on the first or second trial.

Skinner found one very interesting way of magnifying motivation. If he set up his box in such a way that the food pellets arrived every time the bar was pressed, the rats would soon quit working when the food was discontinued. On the other hand, if the machine was set up so that the food pellets came out very irregularly the rat would work much longer when they were cut off. The explanation is that the irregularity of success makes it much more difficult for the rat to discriminate between a situation in which the machine is working and one in which it is not. It follows that partial or irregular success produces much more lasting motivation than uniform success. In training a dog, one does not have to reward the desired behavior every time in order to produce the result. Skinner has suggested that this kind of effect may account for the popularity of slot machines and other gambling devices among human beings.

These results are so interesting that there is a great temptation to assume that they are universal rules. Actually the evidence is not yet sufficiently broad to justify generalization, although it has added greatly to our knowledge of the way in which motivation can be produced by food rewards.

EXTENSION OF PAVLOV'S AND SKINNER'S RESULTS
TO OTHER PATTERNS OF BEHAVIOR

Most of the experimental work on learning in animals has been done in connection with ingestive behavior and the reflexes associated with it. In order to show that these results have real generality, it is necessary to experiment with other major systems of behavior. So far this has been done chiefly with the reflexes and voluntary responses associated with agonistic behavior.

The withdrawal reflex. H. S. Liddell and various co-workers at the Animal Behavior Farm of Cornell University did extensive and important experiments with this reflex. In a typical experiment a dog is put on the same kind of stand used by Pavlov. Instead of food the animal is given a slight electric shock on one foot, which it immediately lifts. If a buzzer is sounded just before the shock, the animal soon begins lifting its foot in response to the sound of the buzzer alone. Comparable experiments have been done with sheep and goats, and the results are very similar to those obtained with the salivary reflex.

Fighting behavior. Most experiments with the effect of learning on this behavior have been done upon the fights between male house mice. Watching these combats, Ginsburg and Allee became interested in the problem of what causes one male to win and another to lose, and the most important factor turned out to be previous experience in fighting. Mice which won an initial fight and continued to win developed very strong habits of fighting and winning, whereas those which were beaten and continued to lose developed strong habits of running away. This suggests that the same principles of association and strength of response apply to agonistic behavior as to ingestive behavior, but before this is accepted the results need to be examined in detail.

When an inexperienced mouse first starts to fight, it shows considerable variability of behavior, nosing the other mouse, jumping away, and finally scratching and biting. If adaptation to the situation is achieved by winning a fight, the mouse will fight earlier and more vigorously at the next opportunity and within a few trials will become a very aggressive and efficient fighter. If it repeatedly fights the same opponent, the reactions of the fighter become less vigorous as the beaten mouse becomes less and less active. Eventually fighting may be reduced to token chasing of the beaten mouse. There is a tendency to reduce activity to the essential motions which cause the beaten mouse to run away, and we can assume that the fighting mouse is discriminating between necessary and unnecessary activity. Furthermore, since the beaten mouse is no longer hurting the victor, the primary stimulus is seldom repeated. When there is a lack of primary stimulation, the response of fighting seems to follow the rule of extinction. Finally, when the fighting mouse is allowed a long period of rest, it will fight much more vigorously for a while, which looks very much like the restoration of the salivary reflex after a period of rest. On the other hand, if the mouse is exposed to a series of different animals which it is allowed to beat, its fighting activities stay at a high level, presumably because it is continuing to receive the primary stimuli. Thus fighting behavior appears to be influenced by learning in the same way as is ingestive behavior.

The principles of learning can also be applied to the problem of controlling fighting. As we saw in the last chapter, mice which are brought up together will frequently not fight for months, even after they become adults. As young mice they have a high threshold of stimulation, and during this time they are likely to form a strong association between not-fighting and the presence of their litter-mates. This habit persists even after they grow up and become more sensitive to stimulation. However, they will still fight strangers, with whom they have no such associations. As stated above, this *passive inhibition* of behavior is the best way to control fighting in human children, since most punishment has the effect of stimulating fighting through pain as well as inhibiting it.

Escape behavior. Different results are obtained with escape behavior. Mice that have been severely defeated in fights with others show reflexes of jumping and squeaking as well as a tendency to run away whenever the victorious mouse approaches. When a harmless mouse is dangled in front of them, they at first show the complete pattern of reflexes and voluntary escape behavior. After a few times the reflexes disappear, but the escape behavior persists for weeks and even months. It is apparent that escape behavior does not extinguish as easily as does ingestive behavior.

The same result occurs in experiments with painful stimulation. R. L. Solomon placed a dog in a box with an electrified floor. A buzzer was sounded just before the electric current was turned on. The dog received a shock, but could avoid this by quickly jumping to the other side of the box as soon as it heard the buzzer. Voluntary as well as reflex behavior was involved in this situation. The dog quickly learned to jump, and then the current was turned off. From the results with other types of behavior, we would expect that the dog would soon stop responding to the buzzer. However, it went on jumping to the sound of the buzzer for hundreds of times, without the primary stimulus ever being given again. We are reminded of trappers' reports concerning the behavior of wolves. Once a wolf has been caught in a particular trap and escaped, it will never approach such a trap again.

One explanation is that avoidance behavior prevents discrimination. If a dog is held in the box so that he cannot escape by jumping and discovers that the electric current no longer comes on, he will modify his behavior rather quickly. Nevertheless, unless discrimination is forced, conditional escape behavior tends to be very long-lasting compared with a conditional food response. We can conclude that the emotional responses concerned with fear are so organized physiologically that they do not extinguish readily, which conforms to the observed fact that fear responses in animals are extraordinarily persistent once they are developed.

Sexual behavior. Experimenting with the effects of castration, Lester Aronson of the American Museum of Natural History found that an inexperienced male cat showed little or no response to females after castration. However, if an adult had the experience of even one successful copulation before the operation, his responses to females persisted for months afterward. In this kind of behavior the effect of a single reinforcing experience is extraordinarily persistent and produces much more of an effect than a similar reinforcement with food. Here again we can conclude that although the general principles of learning apply to all systems of behavior, the underlying motivation and physiological mechanisms may cause profound modification of the results in any practical situation. Similar deviations from the model based on ingestive behavior and its physiology should be expected when the processes of learning are experimentally tested on other systems of behavior, such as care-giving and allelomimetic behavior. At the very least, the systematic exploration of learning should produce a list of primary stimuli that can be used as reinforcers for the control of behavior in both human and nonhuman animals.

THE PRINCIPLES OF LEARNING AS A BASIS
FOR PROBLEM-SOLVING

Almost all animals which move rapidly have to deal with the problem of getting around or over physical barriers. Consequently the barrier test has been a favorite method with those psychologists who have tried to compare learning abilities in various members of the animal kingdom. The type of barrier must be suited to the patterns of behavior peculiar to the species, or the animal has great difficulty in making adjustments. Rats do very well in tunnel-like mazes which resemble their own pathways, but sheep put in a similar apparatus become completely terrified and show little or no evidence of learning. Wild sheep have their natural habitat in open plains and meadows, usually near cliffs to which they can escape. It is possible that a barrier problem which resembled rocky cliffs would be a better test of their abilities.

Learning to run through a maze. A great variety of rat mazes have been used by animal psychologists, starting with extremely complicated patterns like Chinese puzzles or the Hampton Court Maze. However, these have gradually been reduced to simpler and simpler forms until nothing is left but a succession of T- or Y-shaped compartments which lead into each other in random fashion.

If we place a rat in a maze for the first time, its behavior goes through the following steps. At first there is a great deal of exploratory behavior. The rat carefully inspects all the passages, including the blind alleys. Eventually it comes to the end of the

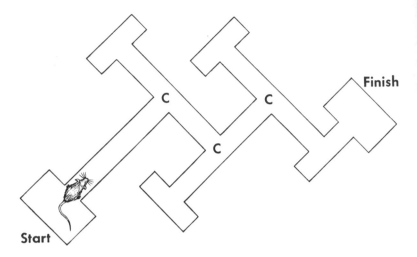

Fig. 15. Simple T maze of two units and three choice points (c). Note that the animal cannot see the end of any blind alley from a choice point. The correct turns here occur in a simple *right-left-right-left* order. With more units connected in a more complicated sequence, such a maze may be made very complex. (Original)

maze where there is a piece of food for it to eat. At this point we take the rat out and put it back in its cage before giving it a new trial. The next time the rat runs, it goes down somewhat fewer blind alleys than before, and with each repetition it tends to take a more and more direct path. Along with this there is a tendency to increase the speed and to persist in the face of difficulties, indicating that motivation is becoming stronger and stronger.

Such behavior is quite different from that required by the Pavlovian experiments or the Skinner box. It reminds us much more of the stentor which varies its responses to carmine particles until some kind of adaptation is produced. Can the rat's behavior be explained in terms of the simple processes of association, or is some entirely new process of learning involved?

Despite the variable responses of the rat, it would appear that the simple processes of learning may be taking place. The rat tends to associate running through the maze with the reward of food. This is quite similar to the rat in the Skinner box which associates bar-pressing with food. By varying its behavior on various trials, the rat is able to discriminate between activities which will lead him to the food and those which will simply cause him to run down blind alleys. The responses which lead to success become more strongly motivated, and those which lead to failure are gradually extinguished. Thus most of the rat's behavior can be explained in terms of positive and negative association.

Generalization may also be important under certain conditions. If the maze is constructed so that it presents a large number of problems that have no relationship to each other, the rat, by varying his behavior, is about as likely to discover a solution to one of them on one trial as another. Since there are more unsolved problems in the early part of his learning, he is more likely to make greater progress in the early part of the test than later, and the result is something like the typical learning curve: an initial rapid rise followed by a much slower rise as the rat reaches a limit of performance. On the other hand, if there is some general resemblance between the problems, as in a maze with alternate left and right turns, the rat may generalize from one problem to all the rest and achieve complete success in one or two trials thereafter. This process of generalization is similar to what is called "hypothesis formation" in more complex learning tasks for human subjects.

One aspect of behavior which the maze emphasizes, and the methods of Pavlov and Skinner tend to prevent, is variability. On the first trial the rat does a large number of different things, and it continues to vary its behavior in subsequent trials. In fact this is the only way in which it could improve its performance. It is not until a relatively late stage in learning that variability is cut down and the behavior becomes a simple and almost automatic habit. Even then the animal never does things in exactly the same way.

Variability versus habit formation. There appear to be at least two basic processes involved in the improvement of adaptation to a particular situation: the associational learning process of habit formation and the tendency toward variability. In any practical problem these appear to work against each other to produce an unstable balance. If a rat is run through a maze several times in rapid succession, habits are quickly made strong, and it tends to repeat mistakes over and over again and to associate these with success. If the rat is given a longer period between trials, variability is increased, and the errors tend to decrease more rapidly. If the period between trials is extended to a matter of days, the associations become very weak, and the rat learns slowly because errors are not remembered. Translating this into practical language, either too much or too little practice produces a slow rate of improvement. We can legitimately ask whether this conclusion applies to problems involving the improvement of human skills. The small girl who plays her piano piece over and over and makes the same mistake each time is probably practicing too often.

We can also conclude that learning to run a maze can be explained by simple Pavlovian principles combined with another basic principle, that of *variability of behavior*. This does not cover all the possibilities of learning and adaptation, and consideration of these more complex types of adjustment will be reserved for another chapter.

LEARNING IN THE ANIMAL KINGDOM

A great many attempts have been made to find out how universal the various phenomena of learning may be, and particularly whether any sort of true learning exists in the lower animals. The available evidence has been summarized in several special volumes, and the reader is referred to these if he is interested in the problem. The evidence is difficult to interpret for many reasons. One is that experimenters have often tried to measure learning by techniques which are unsuitable to the behavioral capacities of the animals with which they work. Cats have been tested with puzzle boxes, in spite of the fact that they are very poor on manipulative apparatus compared to most primates and some other carnivores like raccoons. Mazes have been used for a large variety of animals, some of which adapt well to them and others not. Even the earthworm has been measured in a simple T maze in which it is required to crawl along a surface and turn either right or left, in spite of the fact that earthworms chiefly live in holes, move vertically, and almost never expose themselves by moving along the surface of the ground. Earthworms learn slowly in this situation, but this conclusion may be unfair to the earthworm.

Another difficulty lies in the poorly developed sense organs of lower animals. In an attempt to "condition" the behavior of the flatworm Leptoplana, Hovey found that the worm has a very poor light-perceiving apparatus, capable only of distinguishing between light and dark. Leptoplana starts to crawl whenever the environ-

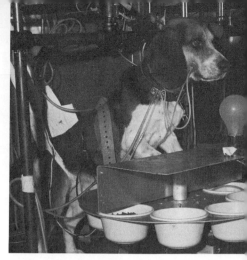

Automated apparatus for operant conditioning of the rat (in white soundproof box). Electronic equipment on the left determines the schedule on which the bar pressing of the rat is reinforced and records the results on paper tape. (Bowling Green University photo)

Beagle in a modern Pavlovian conditioning apparatus. Ducts lead from the dog's salivary glands to a container. Signals are given by the light, and food reinforcement is automatically provided in the feeding dishes. Effect of conditioning on kidney function is being studied. (Photo courtesy of S. A. Corson)

Emotional reactions to separation measured by electronic apparatus. Dr. John L. Fuller measures the heart rate and muscle tension of an African basenji dog after it has been placed alone in a strange room. Most dogs show emotional distress in such situations, some breeds outwardly and some only in their internal reactions. The response forms part of the motivational mechanisms underlying allelomimetic behavior. (Jackson Laboratory photo)

Fighting mice have been much studied experimentally in attempts to identify and analyze the causes of agonistic behavior. Normally, only adult males will fight. The male hormone lowers the threshold for external stimulation. As a result of repeated contacts, the black male on the left has become dominant over the albino, which has adopted a characteristic defense posture, holding out the forefeet and moving only when attacked. (Photo by R. Buchsbaum of mice studied by B. Ginsburg)

Postures of mutual threat in the dog. Each animal snarls while placing its paws on the other's chest and keeping the tail erect. These are Telomian dogs, a breed maintained by the aborigines of Malaysia. Female on left, male on right.

ment changes from dark to light. This may be considered a primary stimulus and its response. The next problem is to find a secondary stimulus which can be associated with the response. Obviously it cannot be anything which the animal sees, because flatworms do not distinguish objects visually. The worm reacts to touch, but this is another primary stimulus which causes it to stop crawling. Hovey finally decided to touch the flatworm lightly on the front each time it was exposed to light and started to crawl. Eventually it stayed still when the light shone upon it, and we can conclude that a new association was made between light and standing still.

The difference between these results and those on higher animals is that a very long period of training was necessary compared to the higher vertebrates, which often make associations within one or two trials. However, this might be the result of working with two primary stimuli.

One of the most fascinating areas of research on learning in recent years has emerged from studies of conditioning the common freshwater planaria, *Dugesia dorotocephala*. These animals have long been laboratory favorites for the demonstration of regeneration. Any high school student can cut planaria in half, and if he leaves them in cool water from an unpolluted natural source to which no chemicals have been added, he will find that each of the two halves soon produces a miniature perfect planaria. The head half grows a new tail and the tail half a new head, the whole body being remodeled to the proper scale.

Like Leptoplana, Dugesia will change its behavior in response to conditioning procedures, although great care must be taken to provide favorable living conditions and avoid varying factors that modify behavior, such as the temperature of the water, the time of day, the presence or absence of slime on the dishes and test apparatus, and the chemical composition of the water. These are extremely sensitive animals, and the unsophisticated experimenter is likely to either kill the animals or obtain highly variable and to him unexplainable results. James McConnell, in a flash of inspiration, tried the experiment of first training planaria, then cutting them in half and training the two halves after they had regenerated. Although during the process of regeneration the nervous system is taken apart and remodeled into a smaller shape, the regenerated worms learned considerably faster than did naïve ones. One would have expected that the head section would perform better than the tail, but this was not the case. Whereas the original worms took 134 trials to reach a criterion of 23 out of 25 positive responses, the regenerated head and tail halves averaged respectively only 40 and 43 trials. This suggested that the basic mechanism for memory lies not in the arrangement of cells to each other but in the chemical compounds carried by the cells, and of the known chemical compounds in nervous tissue that of ribonucleic acid, or RNA, appears to be the most likely.

McConnell then reasoned that it might be possible to transfer a memory trace from one animal to another, and the easiest method was to feed planaria on chunks of planaria that had already been trained. Again McConnell obtained positive results. Worms fed on pieces of trained worms performed better than those fed on pieces of untrained animals. These results appear to be almost incredible, especially since the molecules of RNA are very large and ordinarily must be broken down in order to permit absorption into living cells. The difficulty here is to demonstrate a mechanism by which memory could be transferred in a biochemical fashion, and this is a very difficult thing to do, especially since the process of learning is affected not only by stored information but also by motivation. Another difficulty is that because planaria are so sensitive to various sorts of chemical stimuli it is very difficult for experimenters to repeat each other's results unless they are unusually skilled and thoroughly acquainted with each other's methods. The final explanation of these remarkable experimental findings is still to come. In the meantime they demonstrate that much can be done with even a very simple animal if one knows it thoroughly. With these simple worms, a person with imagination can do experiments that are impossible on the higher animals. Among more complex invertebrates the capacity for learning is better developed. We have already seen that the octopus, with its highly developed eyes and grasping organs, can learn to discriminate in much the same way as a dog.

The generality of the phenomenon of the modification of behavior as a result of experience has been established for the animal kingdom but not the generality of the underlying biochemical processes and physiological mechanisms. Is the modification of behavior in protozoa, which may last only a few minutes, or hours at the best, accomplished by the same mechanisms as the learning of a higher vertebrate like a dog, which can last for several years or throughout its lifetime? Again, most insects are relatively short-lived, but show some sort of learning. For example, digger wasps are able to find the way back to their nests without difficulty.

McConnell has suggested that the same basic biochemical phenomenon underlies the processes of learning in both vertebrates and nonvertebrates. RNA, the same chemical substance that is involved in the transmission of genetic information, is also an important constituent of nerve cells. One of the most promising hypotheses for understanding the physiology of learning lies in this direction. Nevertheless, we still need a systematic exploration of the process of learning in various invertebrate animals before we can arrive at a truly general theory of learning.

In general, experiments with all classes of vertebrates give results essentially similar to those for the rat and dog. For example, fish kept in a pond and fed by throwing in food form an association between splashing and food, as anyone can observe by throwing in a few small pebbles. The fish quickly respond but after a few repeated trials tend to stop rising to the pebbles. Many other experiments support the view that association, discrimination, and generalization are present in these animals. On the basis of present evidence we may conclude that behavior is affected by previous experience in any animals which show behavior. That similar processes of learning are found in all animal nervous systems is a tenable theory, but it has by no means been completely established.

To summarize, this chapter has been concerned with the basic nature of adaptation. This process consists of two parts which somewhat oppose each other. The first is the tendency to vary the response to stimulation, particularly if the first attempt at adaptation fails. Every animal has its repertory of behavior which it runs through when faced with problems of difficult adaptation. The second part is the tendency to modify behavior on the basis of previous experience and so make it less variable. In vertebrate animals at least, the way in which this is done can be stated in definite laws or principles, the most important being that an association is formed between a stimulus and that response which happens to be successful. These principles have been developed chiefly with experiments on the eating behavior of particular mammals. The results can be extended to other types of behavior but are modified in accordance with particular physiological mechanisms behind each type of behavior. The capacity for learning differs a great deal from one animal to another and is closely related to the anatomical and physiological capacities displayed by each. This brings us to another aspect of adaptation which we have implied but not yet specifically stated: that there are differences in all of the internal factors affecting behavior, including the capacity for learning itself, and that these differences are the result of biological heredity.

CHAPTER 6

Heredity and Behavior

The cowbird belongs to the blackbird family but shows certain behavior traits which are very different from its cousins, the red-wings. The adults often spend considerable time in the company of cows, picking up insects around their feet and sometimes sitting on their backs. When the mating season comes, the cowbirds remain together in flocks. The males do not take up territories, and the females build no nests but furtively lay their eggs in the nests of other species. A cowbird egg laid in a warbler's nest hatches out in due time, and the young bird is fed by its foster parents. It spends its entire early life among warblers, but its behavior reflects little of its social environment. Cowbirds brought up in this way never learn the warbler's song, the adults confining their vocal efforts to the usual guttural bubbling of their natural parents. When the young birds are old enough to take care of themselves, they leave the warblers behind and join other cowbirds in their flocks on the pastures and grain fields.

This is a natural experiment on the effect of heredity and training upon behavior. As in the case of the European cuckoo, which has similar habits, heredity seems to be all-important and early environment has little effect on behavior. A cowbird remains a cowbird and a cuckoo a cuckoo, no matter whose nest it has lived in. However, when we do comparable laboratory experiments with other species of birds, we find that the environment also can have important effects upon behavior.

TESTING THE EFFECTS OF HEREDITY ON BEHAVIOR

Modifying the environment. One way to test the effect of heredity is to rear an animal in a strange environment. The European zoologist Konrad Lorenz has used this method with many species of birds. He took young jackdaws from their nests soon after hatching and raised them by hand, entirely apart from other birds. The normal jackdaw is a dark-colored bird which has the reputation of living in steeples and ruins. Like other members of the crow family, it goes in small flocks, makes a lot of noise, and is reputed to be quite intelligent. Lorenz' hand-reared jackdaws attached themselves to him but continued to exhibit many of the typical behavior patterns of jackdaws. If he held any black object, for example a black bathing suit, they would attack his hand in the same way that wild birds attack anything holding a young jackdaw. On the other hand, the tame jackdaws did not show fear of cats and other predators, so that hand-reared birds turned loose in the wild had a poor chance of survival. The adult wild birds make a loud rattling noise as soon as a cat comes near; the young birds take flight and soon learn to make their own associations between predators and escape. This shows that part of the behavior of jackdaws is passed along through learning and that in these animals there is a certain amount of cultural or social inheritance of behavior traits in addition to the biological heredity which sets the general pattern of their behavior.

Similar results can be obtained with mammals. Deer taken as young fawns and raised on the bottle become strongly attached to people and will follow them everywhere begging for food. Though associated with a different species, their basic patterns of adaptive behavior are like those of the wild animals. If you attack or threaten a tame deer, it responds with typical deer behavior, rearing on its hind legs and striking with its front feet, just as a wild deer would try to fight off a predator. If you try to hold a male deer by the antlers, it responds by pushing as deer do when they lock horns and fight. On the other hand, bottle-raised fawns show no fear of people, automobiles, sudden noises, or predators. Their air of calm contrasts greatly with the nervous, flighty attitude of wild animals, and we can conclude that what they fear or do not fear is learned anew from generation to generation.

Bottle-reared domestic sheep act like their parents but direct all their behavior toward people. They follow the person that feeds them and baa plaintively when left alone with other sheep. Unlike sheep reared in a flock, they show no fear of dogs and are the most likely victims if the group is attacked. Even greater indications of social heredity are found in wild mountain sheep. According to historical records, the bighorn sheep of the West formerly migrated each year from the high mountain pastures in the summer to low-lying barren buttes in the winter, sometimes traveling a distance of two hundred miles. Since the plains have been fenced, the mountain sheep no longer follow the migration routes and remain in the mountains the year around.

As we ordinarily see them, dogs respond almost continuously to human behavior. In this case it is possible to do the opposite kind of experiment and separate the dogs from their human environment. At our laboratory we wished to find out how much of their behavior was native to dogs and how much was the result of the human environment. We placed groups of adult dogs in large fields where they could be watched apart from human beings and found that they reacted toward each other with the same basic behavior patterns with which they responded to people. They wagged their tails at each other, growled and barked, and fawned on any animal which was in possession of food. When all these behavior patterns were written down and analyzed, they were found to be essentially the same as those exhibited by their wild ancestors, the wolves. Puppies born to these animals and kept out of contact with people showed the same behavior as the adults except that they were extremely wild and fearful toward people. The puppies literally went back to the wild in one generation. They apparently have an inherited tendency to develop fears toward strangers. As we might expect from the lasting nature of learned fears, once the puppies had developed timidity the friendly attitude of their parents as briefly observed in later life had little effect in overcoming it. Every puppy has to learn to fear or not to fear, but in this kind of experiment, it has little opportunity to learn this from its elders.

All the evidence of this kind leads to two conclusions: First, no matter how great the change in environment, most of the basic patterns of adaptive behavior do not change. As we might expect from the information in the previous chapter, the sensory and motor equipment which an animal has remains the same no matter what the environment, and the animal uses it in the ways which are most effective. No matter how long a deer associates with man, it never uses its hoofs like fingers. Basic patterns of behavior can be suppressed by training, as people punish dogs for eliminating in the wrong places. Yet the pattern of elimination remains the same, with leg-lifting and the tendency to mark certain scent posts.

The second conclusion is that the behavior of many highly social species of birds and mammals is strongly affected by learning and consequently by some degree of cultural inheritance. There seems to be almost none of this in cowbirds and cuckoos, but a considerable amount in jackdaws. Neither a jackdaw nor a deer has much chance of survival under wild conditions if it is deprived of its opportunity to learn from its natural social and biological environment.

We can conclude that basic patterns of adaptive behavior are little changed by large changes in the environment and hence are greatly affected by heredity. At the same time, the individuals and objects toward which these patterns are directed are greatly influenced by individual and social learning. Finally, the degree to which behavior is controlled by heredity varies greatly from species to species.

Comparisons between species. A major problem of understanding the behavior of domestic dogs involves determining their wild ancestry, and one way to obtain evidence concerning it is to compare their behavior patterns with those of various wild species also belonging to the genus *Canis,* such as wolves, coyotes and jackals. We found that we could identify more than sixty different behavior patterns in the dog and that almost all of these could be found in similar forms in wild wolves.

Among these patterns, barking and howling are of unusual interest. By old tradition wolves are not supposed to bark. Actually, barking in some form occurs in all the members of the genus *Canis* and even in more distantly related animals such as foxes. Its basic function in dogs and wolves is that of an alarm signal given when a strange animal approaches the den. Howling, on the other hand, is a signal by which animals that are separated from each other can get into contact. Since people rarely approach wolf dens, there is ordinarily little opportunity to hear them barking, but wolves hunt over large territories and frequently howl to each other while they are separated. On the other hand, people are constantly trespassing on the territories of domestic dogs and consequently hear them bark a great deal. Dogs are seldom separated from their human masters and consequently howl very little, but this behavior pattern can easily be evoked by confining a dog by himself in a strange place. Thus the differences between the two species can be explained in part by different circumstances of living, although the tendency to bark in dogs has undoubtedly been increased by selection because of the value of this alarm signal to human masters.

The quality of howling varies a great deal from dog to dog, but it is basically similar to the long-drawn-out, somewhat melodious sound made by wolves. It is distinctly different from the

V SH

Fig. 16. Wolf greeting. Subordinate wolf (crouching, tail between legs), allows dominant wolf (erect, tail up) to gently bite his muzzle. This behavior pattern is no longer found in most domestic dogs. (From a sketch by Rudolph Schenkel)

mixture of barks, yaps, and howls produced by coyotes, and the somewhat similar complicated vocalization of jackals. This and other evidence, particularly the highly social nature of both dogs and wolves, indicates that dogs were domesticated from wolves alone.

This brings up the problem of how to study the effects of biological heredity upon species differences. The heart of the science of genetics is the study of variation, and the first task is to establish the fact that there is variation between species. As indicated above, most behavior patterns are quite similar in closely related species and they differ chiefly in the frequency of occurrence. In more widely different species, similar behavior patterns become modified in different directions; one would ordinarily never confuse the bark of a fox with the bark of a dog. The problem here is to establish measurements not only of frequency but also of the qualitative differences in the behavior pattern. More rarely there are clear-cut species differences in certain behavior patterns, as in the unique vocalization of the coyote, which is produced by a combination of sounds which are voiced separately in other species.

Once species differences are established, it is tempting to conclude that these are the result of differences in biological heredity. It is, of course, highly probable that this is true in part. However, there is always the possibility that some of the differences are caused by cultural heredity, and one way of testing this possibility is to rear members of two species in the same environment, as when animals are captured at birth and hand-reared, thus being exposed only to human cultural factors. If wolves are reared in this way, they act very much like big, powerful dogs, but they show one behavior pattern that is rarely if ever seen in dogs—that of greeting another individual, and even the human master, by taking his face gently in their jaws. This behavior pattern has apparently dropped out of the dog's repertory, probably because human beings have reacted against its alarming nature and selected against it.

The effects of selection. A second method of testing the effect of heredity is to keep the environment constant and try to change heredity by selection. Hirsch has devised an ingenious method for selecting for behavioral differences in the fruit fly Drosophila. These animals generally show a tendency when in bottles to crawl upward, showing negative geotropism. Hirsch devised a maze with many choice points at which a fly could either crawl upward or downward. The experimenter could put the flies in the start of the maze one day and come back the next to find the flies distributed in several bottles, ranging from a high bottle which held flies that always turned upward and a low bottle holding those that had always turned downward. These two populations were taken as the parents of the next generation and the whole process repeated generation after generation. The two populations soon began to give distinctly different results, showing that even such apparently stable behavior as a tropism is affected by genetic variation.

All domestic animals have been subjected to selection of behavior traits, and one of the most recent examples is the Norway rat. Wild rats can be tamed considerably by catching them young and giving them frequent handling. However, if they are merely raised in the laboratory in cages, they are so wild and fierce that they have to be handled with tongs and gloves. By contrast, the tame strains of albino rats are so gentle that any stranger can reach into the cage and pick them up by the tail without fear of being bitten.

These effects were produced without any conscious selection, but similar results can be achieved by direct experiment. Rats raised in cages are usually fearful and nervous when placed on a large bare table and show this by urinating and defecating. Some individuals, however, are stolid and unafraid in this situation. Selection for the two types over seven or eight generations produces distinct strains, one fearful and emotional, the other not. This and other experiments show that selection can strongly affect the various patterns of agonistic behavior in the rat. Both the external behavior and the accompanying internal physiological and emotional reactions are changed by heredity.

In most domestic animals it is desirable to have strains which show little agonistic behavior, being neither fearful nor savage. Certain breeds of domestic chickens have been selected for other purposes. Cockfighting is no longer a legitimate sport in most parts of the world, but the breed of game chickens still survives from the days when they were carefully selected and bred for their willingness to fight to the death. The cocks are so ferocious that they cannot be used for experimental purposes, but the milder fights of hens have been tested in the laboratory. As might be expected, the game hens almost always win fights against other breeds, even when matched with much larger birds.

Some dog breeds have been selected in the past for their ability to fight. All of the terriers are fighting breeds. They are easily excited and relatively insensitive to pain, with tough skin on their necks and shoulders and strong jaws and teeth. On the other hand, many breeds have been selected for the opposite kind of behavior. Hounds work in packs to find game, and fighting is highly undesirable. Beagles get along peacefully together, and even strange adult males can be put in the same kennel with little more than some growling and barking.

We conclude that selection can modify behavior by changing heredity. It frequently has its primary effect upon physiological and emotional reactions, although there are many cases in which behavior has been modified through changes in anatomy. The waddling dachshund and the swift greyhound are members of the same species, and their difference in speed is largely produced by the different lengths of their leg bones. Dogs are the species in which the most extensive modifications of behavior have been produced, and it is interesting to note that selection in dogs has acted by modifying the occurrence of patterns of behavior which are present in all members of the species but has not produced anything essentially new.

Cross-fostering experiments. Selection gives strong positive results when it is applied to behavior, and there is every evidence that it alters biological heredity. However, it is also possible for parents in the strains being selected to pass along their behavior in other than biological ways. Cross-fostering experiments, in which the young are exchanged between two strains, are an important test for cultural versus biological heredity.

The results usually confirm the role of biological heredity. Inbred strains of mice differ considerably in their fighting behavior. Hungry mice of the BALB/c strain eat peacefully side by side off the same pellet of food, whereas mice of the C57BL/10 strain actively compete for a single pellet. When the two strains were cross-fostered at birth, the young BALB/c mice did not take up the active habits of their foster parents and thus remained true to heredity. The mice of the other strain likewise acted in accordance with their heredity and attempted to take food away from their peaceful foster parents. In this situation the peaceful parents began to compete with their belligerent adopted offspring. Instead of showing cultural inheritance from parent to offspring, the experiment showed that behavior controlled by the biological heredity of the offspring affected that of the parents.

Hybridization between species. Species differences within the genus *Canis* would make an ideal subject for a hybridization experiment, as all crosses that have heretofore been tried between wolves, coyotes, jackals and dogs have produced fertile offspring.

However, these experiments remain to be done. In an entirely different group of animals, the small African parrots of the genus *Agapornis*, commonly known as lovebirds, William Dilger and his students have studied the effect of hybridization on behavior. Females belonging to the species *roseicollis* build their nests in a characteristic and remarkable way. They first cut straight strips of material off large leaves by making a rapid series of bites along the leaf, and then tear off the strips so that each is approximately 12 cm. long. After cutting a strip, the female tucks it under the feathers of her back and rump, securing it by shoving and hooking it with her beak. When she has accumulated several strips she flies off to a hole in a tree and builds a cup-shaped nest out of the strips. About half the strips usually fall out on the way. Females of another species, *fisheri*, cut nesting material in the same manner, but always carry it one piece at a time in their bills. Thus *fisheri* females make many more trips than *roseicollis*. Perhaps the latter species has a selective advantage in that its behavior does not attract so much attention to the nest by possible predators. However, both species do survive, and there are no data regarding the possible survival value of the behavior.

Dilger succeeded in getting the birds of the two species to mate and produce young. These F$_1$ hybrid females cut strips off leaves like both their parents and make the initial movement of tucking the strips under their feathers. However, they never secure the strips by additional movements and hence the strips always fall out. On about 6 or 8 percent of her trials, a hybrid female will carry a strip of material in her beak, and as she goes through successive mating periods she gradually abandons the tucking behavior and carries more and more material in her beak. However, it takes about three years, each including four breeding periods, before her efficiency approaches that of the *fisheri* parents. This contrasts with the rapidity with which lovebirds learn to modify other activities, such as opening the catch on a cage door, which can be learned in one trial.

From the viewpoint of genetics, it can only be said that the complex behavior pattern of the *roseicollis* female is not inherited as a unit. What does seem to be inherited is the tendency to coordinate and organize the movements composing the pattern in a particular way. Even these are not initially performed in a stereotyped fashion, the purebred birds initially varying their behavior considerably and only approaching a stereotyped pattern after several successive mating seasons, suggesting that the final organization of behavior is a process of learning and habit formation. Unfortunately, crosses with F_1 hybrids are sterile and more extensive genetic experiments are impossible.

Cross-breeding within a species. This is very similar to hybridization between species, except that the different populations are strains or breeds, and that the ways in which crosses can be made are virtually unlimited. The critical test of the theory that behavior is affected by biological heredity is a cross-breeding experiment to determine what controlling factors are passed along from generation to generation. The environment is kept constant and heredity is varied by making a Mendelian cross in which the first-generation offspring are bred together and also crossed back to the parents in every possible combination.

It is the dream of every geneticist to carry out such an experiment on dogs. These animals are a veritable gold mine of hereditary traits, having more variability in both form and behavior than any other domestic species, including man. Dr. John L. Fuller and I were fortunate in having the opportunity to perform a Mendelian experiment with these animals at the Jackson Laboratory in Bar Harbor, Maine. We reared puppies in six sunny nursery rooms under standard conditions of feeding and training. Every four months we moved them outside and began with a new group of six litters. With the aid of four research assistants, we gave each puppy a new type of test or training each week throughout most of the first year of life, making every effort to treat all puppies exactly alike and give them the same sort of tests at the same time in life. The general result was to set up a sort of "school for dogs," with ideal educational conditions of small classes and individual attention for each canine pupil.

We first tested several pure breeds. Of these, the African barkless dog, or basenji, and the cocker spaniel were selected as differing in the largest number of behavior traits. A cross-breeding program was set up and reciprocal crosses were made to test for sex-linked heredity and also to test for possible cultural inheritance and other effects of maternal environment. In the latter case, the first-generation hybrids brought up by a basenji mother should be different from those reared by a cocker. Whenever possible the mother and litter-mates were excluded from the test so

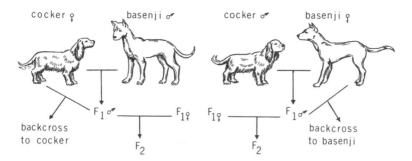

cocker ♀   basenji ♂   cocker ♂   basenji ♀

F₁♂ ———— F₁♀   F₁♀ ———— F₁♂

backcross
to cocker
F₂   F₂

backcross
to basenji

Fig. 17. Mendelian cross testing the effects of heredity on the behavior of dogs. The reciprocal cross (male of one breed by female of the other, and vice versa) tests possible effects of sex-linked inheritance and of the maternal environment. The F_1 (first generation) hybrid males are crossed back to their pure-bred mothers, so that the backcross and F_1 generations are raised in the same maternal environment, and all differences ought to be due to heredity. The F_2 (second generation) hybrids should show maximum variability due to heredity. (Original)

that the puppies could not learn anything by example. We crossed the first generation back to mothers of both strains so that an F_1 generation, which should contain little variable heredity, could be compared with backcrosses having a maximum amount of genetic variability but reared in the same maternal environment.

If heredity has an effect on behavior, the behavior of the crosses should be variable where heredity is variable (in the backcross and F_2 generations) and uniform where it is uniform (in the F_1 generation). We tested a great many traits and got many different results, but the following examples are typical.

One of the most interesting traits of dogs is the tendency to develop timidity of strangers. Young wolf cubs are extremely wild and attempt to bite and struggle as soon as they are picked up. The puppies of most dog breeds still show some of this behavior, but it can often be overcome by a few minutes of handling. Even when the puppies are several weeks old, it takes only a few days to get them used to human company. Under our nursery conditions the puppies were left almost undisturbed until they were five weeks old. At this time we gave them the test described in chapter 2: walking toward them, walking away, holding out our hands, and petting them, all in a routine and standardized way. Everything the puppies did was marked on a checklist. When the test was complete, we counted up all the times when the puppies ran away or made some other kind of fearful response. These we graded so that a puppy that ran away to the corner of the room and hid received a higher score than one which simply ran out of reach. We found that our strain of African basenjis at five weeks started with a much higher avoidance score than did the cocker spaniels. Two weeks later they were almost as tame as the cockers; so we analyzed the results at five weeks for possible inheritance of the trait.

The first-generation hybrids were almost exactly like their basenji parents in their initial timidity scores, which makes it look as if wariness were a dominant trait. The backcross to the basenji parent was also like the basenjis, but the backcross to the cockers included animals some of which were as tame as the cockers and others of which were as wild as the basenjis. As we would expect from the theory of Mendelian heredity, this group was the most variable of all. When we calculated the relative number of tame and wild individuals, it seemed probable that the variability could be accounted for by two genetic factors.

In the above case we are dealing with a wild trait which has been diminished by selection. Other native traits have been strengthened by selective breeding. Anyone who is acquainted with cocker spaniels has noticed the tendency of these animals to crouch or lie flat on their bellies, particularly if they are approached in a threatening manner. This trait has an interesting history. During the Middle Ages birds were frequently hunted with a net. Spaniels were used to find the birds and were taught to crouch as soon as they found them, so that the net could be thrown over both and would entangle the birds as they flew up. Dogs which learned this trait easily were favored as breeders. The whole process was called "setting birds for the net," and a dog which did it was called a "setter." When birds began to be hunted with shotguns, the setting was no longer useful, and dogs which "pointed" or stood still while watching the birds were more helpful. The modern setters are really pointers, and the trait has survived in cocker spaniels only because there has been no selection against it.

In our cross-breeding studies with the African basenjis, the crouching trait came out in many different situations. Like avoidance, crouching in response to a threat is a reaction common to

Fig. 18. Setting birds for the net. *Above,* having found the birds, the spaniel was trained to drop flat on the ground while men came up behind him and threw a net over both birds and dog. (Drawn after Blome, *The Gentleman's Recreation,* London, 1686. Reproduced also in Plate 90 in E. C. Ash, *Dogs, Their History and Development.* Used by permission of Ernest Benn Ltd.) *Below,* spaniels were selected for their ability to learn to crouch. This tendency still survives in many modern cocker spaniels, which will drop flat at a threatening gesture. (Original)

all dogs and shows up strongly in young puppies. We found that when we weighed our puppies there was at first little difference between them, all having some tendency to crouch, sit, or lie down on the scales. As they grew older, the basenjis began to stand on their feet and be active, while more than 60 percent of the cockers would sit down when placed on the weighing platform.

The first-generation hybrids were intermediate between the pure breeds, and when they were backcrossed to the basenjis, the behavior of their offspring was closer to the latter. The backcrosses to the cocker parents were very similar to the pure cockers, and the second-generation hybrids showed a mixture of the two types. Evidently the trait is strongly but not completely controlled by heredity. The ratios indicate that there are probably at least two genetic factors involved. Both cockers and basenjis have inherited the ability to learn to sit, but the stronger tendency of the cockers causes about 65 percent of them to learn from a mild kind of training which does not affect the basenjis at all.

When all the results were put together, it appeared that in most cases more than one gene affected the inheritance of a particular behavioral characteristic. Since there were many different behavior traits, we could conclude that the cockers and basenjis differed by a large number of hereditary factors or genes.

We have described only two of the dozens of traits that were measured, and there were all sorts of other interesting results. For example, we might have expected that the early behavior of each puppy would be quite uniform and consistent, being affected mostly by heredity and very little by the environment. The result was just the opposite. The early behavior of the puppies was extremely variable, and it was only after several weeks of repetition and training that behavior became consistent. In tests where extensive training was involved, it was always true that there was greater variability at first than there was after habit formation had taken place. We may conclude that heredity, while it may limit the behavioral capacities of an animal, does not put behavior in a straitjacket. Rather, it is habit formation which tends to make behavior consistent and invariable. Heredity can modify either variability or habit formation and can tip the balance between them.

We were also impressed by the way in which small environmental factors could interfere with an experiment, even when every effort was made to keep conditions constant. A dog running through a maze might stop for several seconds to investigate and pick up a small chip which had somehow fallen on the floor. The result was to add variability to the test scores for any breed, and in many cases the scores would fall into a series of rough normal curves. The average for each breed was different, but there was always some overlap between them. Among the basenji puppies there were a few which were friendly and fawning like cockers from the very first.

We found that it was extremely difficult to set up a test which would measure only one ability. Basenjis are excellent jumpers and climbers, and we invented a box-climbing test to measure this. Food was placed on the top of boxes of varying heights, and the basenjis had to climb or jump to get it. However, they were quite wary of strange objects in their pens and were not strongly interested in food. They approached the boxes rather hesitantly and did not make unusually good scores. By contrast, cockers are not naturally gifted in jumping or climbing. On the other hand, they were not wary of the boxes and were strongly interested in food. They would back off at a distance, run full tilt at the boxes, and scramble up with such energy that they often got the food quicker than did the basenjis with their superior ability. In any practical situation a dog does not necessarily limit himself to one hereditary capacity but attempts to organize all the capacities which he has to solve the problem.

Many dog owners ask the question: Which breed is most intelligent? The answer that we found was that all breeds would do equally well provided they could be equally motivated, did not show emotional reactions that interfered with the test, and did not have sensory or motor defects that interfered with performance. At one time we found that many of the dogs were doing either well or badly in all the problem-solving tests, but it turned out that this correlation was caused by the fact that the basenjis and some of their descendants showed a fear of strange apparatus that made them relatively poor performers.

The only thing that looked like a purely "intellectual" trait was the tendency of beagles not to form fixed habits. This made it difficult to train them to do routine tasks, such as going into a cage, but led them in the long run to become top performers in a maze test. This and much other evidence on the effects of selection on problem-solving behavior in mammals indicates that the variable heredity that is most easily affected by selection is that affecting emotional rather than purely intellectual or cognitive performance.

Since behavior is so variable and can be so greatly modified by learning and functional organization, we wonder how it is possible for heredity to produce really important behavioral differences. One of the answers seems to be the magnifying effect of a threshold. In a test set up so that one dog cannot do it and another can barely succeed, the initial difference in hereditary ability may not be great. However, the dog which fails soon stops trying, while the one which succeeds becomes more highly motivated with each success. It keeps on trying and succeeding at more and more complicated problems so that in the end the original hereditary difference has been immensely magnified.

Another point is that motivation leads to practice. When a beagle is put out into a field, it is excited by the new smells and sights, whether there is game present or not, and travels over every yard of ground. A fox terrier in the same surroundings is completely bored unless it can hear something or see a moving object which it can chase. The result is that the beagle soon accumulates an immense amount of practice in hunting, which adds to the difference in original ability.

Finally, the different breeds of dogs seem to differ in their range of general adaptability. In the standard environment which we set up, most of the tests were done by the dogs independently, in order to avoid the possibility of receiving unconscious help from the experimenters. All of the dogs were given human handling but nothing like the amount they would get as household pets. Under these conditions all of the dogs got along well physically, and when it came to passing the tests, the hunting breeds, like the cockers and beagles, were excellent performers. However, working dogs like the Shetland sheep dog, which has been selected for its ability to develop a very close relationship with human beings and work well under direction, showed up relatively badly on many of the tests. They did not develop confidence under these conditions and often seemed to be waiting for someone to tell them what to do rather than going ahead and solving the problem by themselves. The possession of a wide or narrow range of behavioral adaptability seems to be at least in part a result of heredity.

To summarize, the cross-breeding experiments show that under carefully controlled environmental conditions some behavior traits segregate in accordance with the laws of Mendelian heredity, although even under these conditions there is considerable environmental variability. Most traits are affected by more than one hereditary factor or gene. Heredity does not destroy an animal's capacity for variability of behavior, though it may limit its capacity for successful adaptation. Animals are capable of combining their native capacities in a variety of ways to meet complex problems. One of the most important ways in which heredity affects behavior is to increase or decrease the ease with which an animal can be stimulated or motivated. This in turn has an enormous effect on the ease with which an animal can learn.

HOW HEREDITY PRODUCES ITS EFFECTS

From a scientific point of view, it is not enough to show that differences in behavior can be transmitted from parent to offspring in the same way that the genes or hereditary units are passed along through the chromosomes. We must also demonstrate the paths through which genes produce chemical and physiological changes which in turn affect behavior. These problems are often difficult, but there are certain well-known ways in which hereditary mechanisms affect behavior.

The effects of anatomy. In this and a previous chapter we discussed numerous examples of anatomical differences in sensory and motor organs which have a decided effect upon behavior. Where the genes can affect growth directly, there is no difficulty in explaining their ultimate effect upon behavior.

One of the first studies on heredity and behavior was described by the famous animal psychologist Yerkes in his book *The Dancing Mouse.* Mice normally live quietly in their cages, but the waltzing mouse, which has been known for centuries in Japan, runs around in circles in a sort of gentle lope whenever it is stimulated. Since Yerkes' study, more than thirty different strains of mice with defects of this type have been discovered—circlers, shakers, and jerkers. Each trait is caused by a single gene, and those strains which have been thoroughly studied show various sorts of anatomical defects in the nervous system which account for the behavior.

One of the normal traits for which the hereditary mechanism is well known is that of sex, which affects both anatomy and behavior. The original difference between the sexes is explained in terms of chromosomes. A great many species are like the fruit fly Drosophila, which has a special pair of chromosomes for sex determination. Females always have two X chromosomes (as they were originally called when no one knew their function), but males have one X chromosome plus a slightly different Y chromosome, which is almost empty of genes. Bees and other members of the insect order Hymenoptera have another kind of sex-determining mechanism. Queen bees can lay either unfertilized eggs with the haploid or "half" number of chromosomes, which develop into males, or fertilized eggs with the diploid or full number, which develop into females. Diploid males are also produced, but these are invariably killed before maturity by other members of the colony.

In insects, accidents to the chromosomes sometimes happen early in the development of an egg, so that an X chromosome may be lost out of one cell in an embryo female fruit fly; or perhaps an egg which started out to be a male bee or wasp later has one cell which fails to divide when the chromosomes divide. The result is a gynandromorph, an animal whose body is part male and part female in appearance. Sometimes the front half may be male and the back half, including the sex organs, may be female. Which half of the body will control the sexual behavior? The answer is that the behavior goes according to the head end, despite a completely inappropriate set of sex organs in the rear.

In the tiny wasp Habrobracon, which is parasitic on mealworms, there is a clear difference in male and female sex behavior. When introduced to females, the males run around excitedly flipping their wings, while the females remain passive. In addition, the males are indifferent to mealworms, but the females sting mealworms, suck their juices, and subsequently deposit eggs in them. The geneticist P. W. Whiting collected some fifty gynandromorphs of this species and tested their behavior with females and caterpillars. The heads and bodies of the wasps were divided in all sorts of ways. When the head was male, the wasps flipped their wings at females and were indifferent to caterpillars; but wasps with female heads were indifferent to females and went through the motions of stinging the caterpillars, even though they had no stings. Some wasps with mixed male and female heads acted like males and others like females. One of these acted as if it had its "wires crossed" and flipped its wings at the caterpillars instead of stinging them.

Another parasitic wasp, Nemeritis, normally never produces anything but females. The developing eggs never reduce the number of chromosomes, and the resulting females lay their eggs on mealworms generation after generation without help from a male. Treating the eggs with X-rays produces chromosome abnormalities, and once in a while a perfectly formed male wasp is produced. However, it never shows any sexual behavior, nor do the females respond to it. The species still has the capacity to produce the two types of anatomy but has completely lost the sexual behavior which goes with it. Heredity may affect behavior through anatomy in these cases, but if so it is the detailed anatomy of the central nervous system which is important rather than the form of the motor organs. The anatomy of the sex organs affects the gynandromorph only to the extent that it makes successful mating behavior impossible. We conclude that heredity can produce important effects on behavior by modifying the growth and development of various parts of the body, but that anatomy may not be the only hereditary determinant of behavior.

Physiological effects of heredity. Sex determination in vertebrate animals is produced in a somewhat different way. The distribution of sex chromosomes determines whether a developing embryo will have an ovary or a testis. These primary sex organs in turn produce female or male hormones which affect both anatomy and behavior. This can be tested by castration experiments. Poultry breeders castrate male cockerels with the result that they develop into fat capons, showing little comb development or sexual behavior. If the same operation is performed upon a female chick, the result is frequently not a capon but a bird with comb, wattles, and tail feathers like those of a cock. Later it begins to look more like a hen again. The explanation is that in birds only the left ovary is highly developed. If the vestigial right ovary is not removed with the left, it becomes enlarged and produces hormones. Its tissue contains both male and female cells and it becomes an ovatestis. With both hormones present, the partially castrated hen takes on some of the characteristics of both sexes. If both ovaries are removed, the pullet changes into a bird almost exactly like a capon.

If a hen is kept to a ripe old age, it eventually stops laying and the left ovary degenerates. When this happens the right ovatestis begins to be active, and the hen may begin to show growth of the comb and wattles and some of the male patterns of behavior. There is at least one fairly authentic case on record of a hen which underwent a complete sex reversal, showed male behavior, and actually fathered a few chicks. Partial sex reversal is common enough that it was well known in the Middle Ages, when crowing hens were considered bewitched and immediately destroyed.

All this experimental evidence indicates that chickens are capable of either male or female patterns of behavior, depending on which hormone is present in high concentration in the blood stream. In chickens the male and female external sex organs are so much alike that the complete mating behavior can be carried out by an animal with altered sex.

The situation is somewhat different in mammals, where the external sex organs are quite unlike in males and females. The male or female hormone causes the early development of either male or female organs, but once these are formed they cannot be reversed later in life. However, a castrate male injected with the female hormone will act like a female and vice versa. As in birds, every individual is capable of either male or female behavior. In fact, it is extremely common for female mammals to mount each other or attempt to mount males during periods of high sexual excitement, as we can observe in a herd of cattle or in other domestic mammals where large numbers of females are confined together. In the female mammal, anatomy has a slightly greater effect on behavior than in birds, since the complete pattern of male behavior is impossible without the external male organs.

As we saw in chapter 4, the sex hormones affect fighting behavior differently from sexual behavior. Fighting is to some extent influenced by motor capacities, since male mammals are usually larger than females, but the most important effects are produced by an early modification of sensitivity to the male hormone, presumably produced directly on the nervous system. Thus the sex hormones not only have activating effects on the behavior of adults, but have organizing effects on the anatomy and nervous systems of the unborn fetus, and in some cases of the new-born animal.

The sex hormones exert similar organizing functions on sexual behavior, mostly through the modification of the secondary sex organs. It has long been known that when male and female twins are born in cattle, the external genital organs of the female are likely to be modified in the male direction, producing a sterile animal that cattle breeders call a freemartin. W. C. Young and his co-workers found that if they injected male hormones into a pregnant monkey bearing a genetically female fetus, she produced a masculinized infant with malelike sex organs which later showed more masculine behavior than normal females.

It is obvious in all of these cases that the anatomy of the motor organs is not the only means through which behavior is affected by heredity and that in both vertebrates and insects other factors are involved, such as the modification of the structure and function of the nervous system in both groups, and the physiology of hormones in vertebrates.

Magnification of genetic differences by a threshold. Fruit flies have long been a favorite research animal with geneticists. These tiny insects show all sorts of variations in eyes, wings, and body form; and hundreds can be raised in a half-pint milk bottle in a period of a few weeks. As we glance at a row of bottles on a shelf, we see that different stocks exhibit differences in behavior as well as anatomy. Some of them readily crawl toward light and others do not. In one experiment we repeatedly crossed red-eyed flies with a white-eyed stock until there was little chance that there was anything but the single genetic difference of eye color between them. When tested with very weak light, the white-eyed flies would crawl when the red-eyed ones would not.

The two kinds of flies obviously had different thresholds of stimulation. At a high intensity of light both kinds crawled toward it, but at a low intensity the difference in thresholds made the difference between a fly which crawled fairly rapidly toward light and one which did not crawl at all. The difference was presumably caused by the fact that the white-eyed flies had no pigment screen in their eyes and hence could respond to weaker light.

Heredity can thus produce important effects on behavior by raising or lowering thresholds. In the case of timidity in dogs described above, all young puppies can be frightened if the stimulus is strong enough. The mild stimulation of ordinary human handling frightens some strains of puppies but not others. Similarly, the cocker spaniels have a very low threshold of stimulation for the crouching reaction, and simply picking them up and placing them on the scale will bring it out. The rats which were selected for their emotional reactions to being placed on an open table have a low threshold for this trait.

This and much other evidence shows that one of the important ways in which heredity affects behavior is by raising or lowering thresholds. The actual physiological differences may not be great, but if the level of stimulation is near the threshold, there may be an absolute difference between animals which react and those which do not.

Biochemical effects on behavior. Professor Dice of the University of Michigan had long been interested in the heredity of the deer mouse. One day he was walking through his colony, absent-mindedly jingling a bunch of keys, and he noticed that some of the mice were rolling over in convulsions as he passed. Investigation soon showed that it was the jingling keys and not his appearance which caused the fits, and since that time other workers have found these audiogenic seizures in many kinds of rodents. The trait is hereditary, and all that is necessary to set it off is to make a high-pitched noise. Door bells and air blasts are just as effective as jingling keys.

One summer a group of scientists interested in the study of behavior were gathered at the Jackson Laboratory. Dr. C. S. Hall decided to survey the pure strains of inbred mice, which had been developed for cancer research, for their susceptibility to sound. Of the first two strains that he tried, the dba strain showed almost 100 percent seizures at thirty days of age, while the C57BL/6 strain showed almost none. The trait was inherited in Mendelian fashion.

Dr. Benson Ginsburg became fascinated with the possibilities of studying the physiological mechanisms through which the genes produced the trait, and careful cross-breeding studies showed that it was controlled by two genes. He then began to test drugs which affect the susceptible animals and found a common clue in the fact that anything which affected the metabolism of the nervous system will either increase or decrease the seizures. The brains of mice with seizures are deficient in two substances which are normally used in metabolism. The deficiency presumably produces the seizures by interfering with the energy supply for normal brain or muscle action. Biochemical differences can be traced to specific areas in the brain.

Early experience also has an effect on the susceptibility to seizures. Other workers, including R. L. Collins and John Fuller, have shown that mice of the C57BL/6 strain will become susceptible if they are primed by exposure to loud sounds as early as thirteen days, the exposure showing a peak effect at eighteen days. Such priming is not necessary to the dba strain.

These experiments show that heredity can affect behavior in other ways than by modifying growth or producing hormones. It can directly affect the normal physiology of an animal, and we have reason to believe that it may affect the physiology of the nervous system itself. Other examples of genetically caused deficiencies of metabolism are found in man. One rare inherited disease, called "phenylpyruvic oligophrenia," is characterized by two defects which are always associated. The victims excrete in their urine large amounts of phenylpyruvic acid, and they are also defective in intelligence. The development of the latter defect can be prevented by early treatment aimed at preventing the accumulation of large quantities of the chemical substance which is destroyed in normal individuals by the usual processes of metabolism.

As we saw in the last chapter, the process of memory formation is almost certainly biochemical in nature, involving nucleic acid in the form of RNA (ribonucleic acid). Since this substance is known to be associated with primary gene action, many authorities believe that the genes are directly involved in the process of learning. At any rate, it is likely that, once the biochemistry of learning is completely known, genes will be discovered that have effects on each part of the process.

Instinct. The whole preceding discussion of the effects of heredity on behavior has been carried on with scarcely a mention of the word "instinct," but the term needs to be discussed from a historical viewpoint. At the present time it is still widely used by those biologists who are principally interested in comparing the behavior of different species, and relatively little used by psychologists, many of whom state that the word has come to have so many misleading connotations that its use should be discontinued.

To understand this disagreement, it is necessary only to remember that the term was originally used before the rediscovery of Mendel's work; that is, before the development of the modern science of genetics.

As used by biologists in the nineteenth century, instinct meant an inherited pattern of behavior. We now know that the only things that can be biologically passed along or inherited from the parents are the nuclei of the egg and sperm cells, which include the genes, plus anything which is passed along in the cytoplasm of the egg. In the case of mammals, the physiology of the pregnant mother has biological effects on the developing egg in her uterus, and her care of the young can directly affect behavior after birth. This last, of course, is cultural rather than biological inheritance. Therefore, from the viewpoint of genetics, there is no possible way in which behavior as such could be biologically inherited through the chromosomes. We can say that biological heredity has in many cases important effects compared to other factors which modify behavior, but that the capacity for behavior has to be developed.

The problem in modern terms is simply this: How can heredity affect behavior? As indicated earlier, heredity can affect the *growth* of sensory, motor, and coordinating organs of animals and so limit what they can do. It can also modify other physiological processes, such as the production of hormones which affect behavior and the physiological processes which go on in the nervous system itself.

Heredity therefore does not create behavior but rather modifies it and helps determine differences between individuals, groups, and species. Neither is behavior created by the processes of learning; rather, these processes change and modify the behavior of individuals and, because they act differentially, also help to determine differences between individuals, groups, and species. In any animal capable of learning (including all vertebrates and probably most invertebrates), both kinds of organizing processes are present and interact with each other. Heredity may even determine differences in learning capacities. The original controversy as to whether particular segments of behavior are inherited or learned thus becomes a dead issue. The word "instinct" still has a legitimate use as a general descriptive term applied to behavior whose variation is largely determined by heredity, but it is no longer a useful analytical tool and should not be employed as an explanation of behavior.

A second problem arises out of the descriptions of behavior that have been collected under the influence of instinct theory, and indeed form the central problem behind this theory. This question is: Why does behavior in a particular species develop so consistently, even allowing for genetic and environmental variations? Ducks always grow up to quack and swim like ducks, even though they may have been reared entirely by a human experimenter. One answer, and a guide for deeper research into the problem, is provided by the concept of organized systems and the correlated theory that an individual is conceived, grows, and lives under the influence of many preorganized systems that guide his development along a relatively narrow and restricted path. Parts of these systems are so fixed and invariable that, like the influence of gravity, which could be measured only in space flight, their influence can be tested only by a series of extraordinary experiments.

The first of these preorganized systems is the genetic one. Not only are the genes of a particular individual organized into chromosomes, but the entire gene pool of a species forms an organized system—variable, balanced, and flexible. The second major system is that of the cytoplasm of the egg in which the genes exist, formed before conception under the influence of the genetic and physiological constitution of the mother. In egg-laying animals, the egg is the important part of the environment before hatching. For a mammal, and other animals which bear their young alive, early development takes place within a third preorganized system, the body of the mother herself, in which environmental conditions for the developing embryo are ordinarily kept constant by the processes of homeostasis. After birth, the young mammal emerges into other preorganized systems. For a highly social animal the system of social organization may be extremely stable, generation after generation. A young lamb is always born into a flock of sheep, and a young wolf is born into a litter of wolf cubs with a mother who provides constant care for the first few weeks of life.

The nonsocial world is also preorganized. The biotic environment forms an ecosystem in which various species of animals and plants maintain themselves in a state of unstable equilibrium, but one which nevertheless ordinarily changes quite slowly. Finally, the physical environment itself is organized on a grand scale. Night follows day, and the seasons succeed each other in a highly regular fashion, even allowing for variations in weather conditions. As was pointed out above, the gravitational component of this system changes scarcely at all.

The result of the interaction of a particular genetic system with all these preorganized systems is thus highly predictable. It is only by deliberately disturbing these systems that one can determine the influence of each. The genetic system can be modified by cross-breeding and selection, and the prenatal maternal environment can be upset by drugs and other external factors. Modifying the cytoplasm of an egg is more difficult, but when fertilized nuclei are transplanted into the cytoplasm of a different species, development either stops or proceeds in an abnormal manner.

It is the last three of these preorganized systems—the social, biotic, and physical—that have been subjected to the greatest degree of experimental modification. It is obvious that an animal changes and adapts its behavior to changes in the physical and biological environment, but only recently have we discovered the profound changes that can take place as a result of modifying the normally stable social environment, as will be seen in the next chapters. It is also these last three systems which are chiefly involved in the organization of behavior through the process of learning. If we are to understand the origin of behavior, we must systematically explore the results of modifying each of these organized systems and the interactions between them.

As a result of the long history of this problem of predictable development, there are several fallacies that persist concerning the relationship between instinct and learning. One is that behavior can be classified as either instinctive (innate) or learned. This fallacy arises out of the concept that behavior can be independently created by either heredity or environment. A more correct concept is that behavior always develops under the influence of both heredity and environmental factors, each of which can modify development.

A second fallacy, arising from the above, is that behavior can be recognized as either instinctive or learned on the basis of variability; that is, instinctive behavior is always stereotyped and learned behavior is not. Actually, the essence of behavioral adaptation is variation. Without altering its behavior, an animal cannot adapt to new situations. In any scientific study of behavior, a basic technique is the measurement of variation, both within and between individuals. The relative amount of variation that can be attributed to hereditary factors (the heritability index) can be calculated for any particular situation, but the results of course depend upon the amounts of genetic and environmental variation that are permitted by the experimenter. Even where we deliberately maximized the effects of heredity in our dog experiments by working with several contrasting breeds under conditions of a standardized and relatively invariable environment, we found that the heritability index rarely went over 50 percent. On the other hand, there is almost no behavior that is unaffected by heredity, unless we restrict our measurements to a genetically uniform population like an inbred strain of mice.

This brings us to a final fallacy which arose out of the typological approach of the older biologists who worked before the rediscovery of Mendelian heredity. This was the assumption that by studying the behavior of a single animal, or even two or three animals, we could discover behavior which is general to the whole species. On the contrary, we now know that if we want to describe the behavior of a species, we must sample an adequately large number of individuals, sampling not only the different genetic subpopulations but also different environmental conditions under which the species lives. The results can then be described in terms of the frequency and variation of particular kinds of behavior. Wherever this has been done with wild species, as for example in the songs of birds, results show not standardized behavior, but wide variations.

Modern scientific problems in this field are concerned with the deeper analysis of the behavior that develops consistently in the members of a given species. In the first place, in a developing animal there has to be some sort of primary behavioral reaction on which learning and environmental factors can operate. As we found in the development of puppy behavior, these primary actions are often quite variable and become fixed only through later habit formation. This leads to the problem of the extent to which behavior is organized as the result of hereditary factors as compared to the extent to which it is organized by the function of the nervous system and the various processes of learning. This matter will be taken up in the next chapter.

Species-specific behavior. In searching for a term that retains some of the meaning but not the undesirable connotations of the word instinct, many scientists use the term *species-specific behavior.* In a narrow definition this term can be used to describe behavior only found in one species. Actually, most behavior patterns are widely found in closely related animals, at least in slightly modified forms, and in general usage the term means behavior that is found in all the members of a given species but is not necessarily confined to that species alone. For this reason, many scientists prefer the terms "species typical" or "species characteristic" instead. All these terms are useful in that their meanings are neutral, not implying that any particular set of causes underlies the behavior. This leaves the experimenter free to analyze the behavior in any way possible.

HEREDITY AND HUMAN BEHAVIOR

We may complete this description of heredity and behavior by applying the results to man. Heredity has been so much used and misused to explain and justify human behavior that it is important to have a clear idea of the ways in which the general conclusions derived from animal experiments apply to human affairs.

There is a tendency in the evolution of social animals to place greater and greater emphasis on cultural inheritance as a determinant of behavior. Jackdaws learn their fears from their elders, and sheep learn to follow each other and use the old trails and migration routes. With the great development of language in human beings, cultural inheritance becomes even more important in relation to biological heredity.

There is also a tendency among animals to evolve toward greater and greater degrees of organization of behavior through learning. Even dogs are able to organize their abilities in different ways to meet new situations, and we have every reason to believe that this ability to organize behavior functionally is highly developed in human beings. It might almost seem that biological heredity no longer has any important effect on human behavior.

This, however, is definitely not the case. There are many genet-

Breeds of dogs selected for various behavior traits and hunting abilities. *From left to right* are the beagle, developed as a scent hound for hunting rabbits; the basenji or African barkless dog, originally used in pursuit of small game; the Shetland sheep dog, related to Scotch collies, selected for their ability to learn to herd sheep; the cocker spaniel, related to bird dogs and used for flushing and retrieving; the wire-haired fox terrier, aggressive and used to attack small game. (Jackson Laboratory photo)

Testing a young puppy for tendency to crouch or lie quietly while being weighed. The puppy is a two-week-old hybrid between a cocker spaniel and a basenji. Crouching is strongly developed in cocker spaniels and is known to be affected by at least two genetic factors. (Jackson Laboratory photo)

Testing social reactions of a puppy to a human handler. Test consists of doing things which people ordinarily do to a puppy, but in a standard and consistent fashion, and keeping a careful record of responses. This test is particularly useful in picking out puppies with a tendency toward avoidance and timidity. (Jackson Laboratory photo)

Test of climbing ability. A dish of food is placed on top of three boxes, and the dog must climb to get it. On successive days the boxes are built higher, making the task more difficult. Basenjis have unusual climbing ability but are timid about approaching the apparatus, so that cocker spaniels, with less ability but more motivation for food, are apt to perform equally well. The dog in the picture is a second-generation cocker-basenji hybrid. (Jackson Laboratory photo)

Nest building behavior in caged lovebirds, *Agapornis roseicollis.* **One of the pair has just thrust a** strip of paper among its tail feathers preparatory to flying to the nest. (Photo courtesy of William C. Dilger)

Puppy confronted by a detour problem. In order to reach the goals of food and contact with the experimenters, the puppy must first go away from them. (Jackson Laboratory photo)

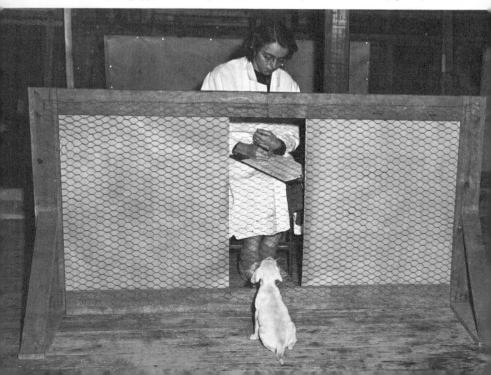

ically produced human traits which drastically affect behavior. For example Huntington's chorea causes a degeneration of the nervous system in middle age. Among normal individuals there are wide differences in sensory and motor abilities which are almost certainly the result of hereditary effects on growth, and these in turn affect behavior. A heavy-set muscular person can usually put the shot better than a tall thin one, whereas the latter can often do better in the high jump. There is probably also a great deal of variability in such traits as sensitivity to emotional stimulation. All this means that human beings are almost infinitely variable in the combinations of genetic traits and capabilities which they may have.

One place where we still cannot generalize from animal to human behavior concerns the effect of hereditary variability on the power to organize behavior, or what we commonly call intelligence. There is plenty of evidence from animal experiments that individual hereditary differences in emotional reactions and in sensory and motor abilities will affect the power to adapt in a particular situation and that defects in the nervous system will cause defects in behavior. But there is still no clear-cut evidence that hereditary differences in the general power to organize behavior exist apart from these. This whole problem will be considered in greater detail in the next chapter.

There are many cases in which heredity may produce important differences between human individuals because of the magnifying effects of a threshold. A hereditary difference in ability amounting to an inch in a high jump may make the difference between a world's champion and an ordinary athlete. A mild case of undiscovered deafness may make the difference between a child who learns well in school and one who is a failure.

At the same time we need to remember that there are no pure breeds among human beings and that differences between individuals are far more important than any differences which have been demonstrated between populations. Heredity contributes to human behavior by increasing its variability, and in the long run this permits a highly complex degree of social organization. There may be four or five different castes in a social insect, each of which shows specific types of behavior. There are two basic biological types of human beings, males and females, but because of variable heredity each can be subdivided into hundreds and thousands of other biological types. This makes possible the division of labor between individuals having a great variety of capabilities. There is a synergistic relationship between social organization, which tolerates and protects a wide range of variant individuals, and genetic variation, which permits a more elaborate and efficient social organization. At the same time, heredity does not prevent us from undertaking common tasks and enjoying common behavior. Since human beings are also very adaptable, they can combine many different sets of abilities to produce the same sort of end result. In order to take full advantage of our variable heredity and flexible behavior, we must be allowed to behave in different ways, and the ideal form of social organization is one which permits a great deal of freedom and tolerance for variability in human behavior.

CHAPTER 7

Intelligence
The Organization of Behavior

For thousands of years poets have praised the beauties of bird song, interpreting it in various ways as the love song of the bird to his mate or a simple expression of physical well-being and joy in the return of spring. The usual poetic rendition of bird song is, "Cheer up! Spring is here!"

It is true that the singing of birds is greatly increased during the spring and early summer, and a camper in the woods at this season is likely to be awakened in pitch darkness at two or three o'clock in the morning by a clamor of different bird sounds. Serious students of bird behavior early discovered that each species has a characteristic song, or at least some characteristic noises, and that these noises are made at certain times and seasons.

It remained for the English naturalist Eliot Howard to put all this information together and discover that in the ordinary song of birds the most musical part of the performance has a very different connotation for birds than it does for man. Among the birds that Howard studied was the English reed bunting, which has many habits like those of the American red-winged blackbird. The bunting does not migrate, but in the early spring the males go to particular spots among the reeds, where they perch in conspicuous places and sing. This is before the females join them, and so it seems likely that the song is not primarily intended as a serenade to delight the female. The spot from which the male bird sings becomes the center of his territory, which he guards against all other male buntings. There can be no doubt that the song itself serves mostly as a notice that this particular bit of territory is already occupied and that any bird which disregards it is in for a fight.

This is only one of many notable examples in the history of

147

animal behavior in which the obvious human interpretation of behavior later turned out to be incorrect. This particular kind of mistake is called anthropomorphism, or interpreting animal behavior in terms of human behavior. In psychological terms it is related to the human mental process of projection. The human observer sees the animal as a human being and finds his own motives reflected in the animal's actions. This is not necessarily false, but it should be remembered that it is only a hypothesis, and every scientist must consider the alternate possibility that the animal may be responding in ways which are entirely its own.

So far in this book the analysis of behavior has dealt with examples of adaptation and causation which were clear and straightforward in interpretation. In any actual case the student of behavior is continually observing things which are not immediately clear and comprehensible, and a series of questions continually arises in his mind. First, what is the animal doing, and what is the adaptive value, if any, of its behavior? One answer for bird song is that it appears to be part of a system of dividing up the breeding grounds into territories. A second question which often arises is: To what is the animal responding? and a third: How is the animal organizing its behavior? Is it employing some kind of reasoning process, or is it merely responding in a mechanical way to a problem which has been facing its species for thousands of years? The answers to these questions are not always simple, and obtaining them usually involves long periods of careful observation and experiment. In general they cover the many phenomena which used to be lumped together as intelligence.

THE ADAPTIVE VALUE OF BEHAVIOR

Tropisms. One of the first experimental techniques applied to animal behavior was to make large changes in the physical environment and observe how animals reacted to them. In many cases the animals responded by either approaching or avoiding the source of the change or stimulus, and these reactions were called *tropisms*. For example, the tendency of laboratory fruit flies to congregate on the light side of their bottles is called "positive phototropism." On the other hand, a silverfish, a tiny wingless insect often found in the vicinity of old books, will scurry away from the light toward a dark corner if it is suddenly uncovered. This is called "negative phototropism." Some writers prefer the term "taxis," in reference to animal behavior, reserving tropism to refer to the turning of plants. Thus the behavior of silverfish can also be called "negative phototaxis."

Reactions of this kind are not necessarily simple and mechanical in nature. When I was interested in investigating the apparent hereditary differences in phototropism in fruit flies, I first tried to set up an experimental situation which would measure it. Every precaution was taken to raise the flies under standard conditions. All the males that were hatched within a few hours of each other were placed in a new bottle and tested for their phototropic reaction a few hours later. This was done by placing them in one end of a long tube with a dark barrier in front of it. After they had had time to settle down completely, the barrier was lifted and a beam of light directed down the tube. Surprisingly enough, the flies for the most part remained perfectly still, and then after a few minutes slowly roused themselves and crawled around in various parts of the tube. Some minutes later when the experiment was repeated, the group of flies immediately started to crawl toward the bright end of the tube as they were expected to do. Sometimes they responded to the light and sometimes they did not. It was only after several days that I realized that heavy trucks were occasionally passing the laboratory and shaking it to its foundations. When this happened the flies crawled rapidly toward the light. After that it was easy to get the flies to respond. Just before the barrier was lifted, the table was mechanically jarred and all of the flies thereupon surged down toward the lighted end.

Seen in this way, the "phototropism" of the Drosophila was no longer a curious mechanical reaction to stimulation. If fruit flies in a state of nature were truly positively phototropic, they would fly toward the sun until they were exhausted, and no flies seem to do this. However, fruit flies often go into cavities of rotten fruit or into dark places like garbage pails. Under these circumstances flying toward the light when the surroundings were jarred would be a simple escape reaction of considerable value to the species.

Among the animals found in the ocean near Woods Hole, Massachusetts, is a species of brittle star. When these agile cousins of the starfish are brought in from the sea bottom and placed in a laboratory tank, they almost immediately crawl into the corners of the aquarium and stay there. They are showing positive "thigmo-

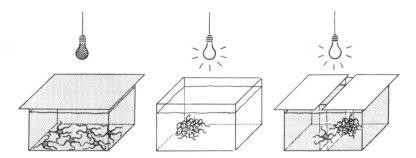

Fig. 19. A "tropism" in brittle stars. Animals are placed in an aquarium under various light conditions. *Left,* after several hours in complete darkness they are scattered at random over the bottom. *Center,* in bright light, the brittle stars huddle together in one corner of the aquarium. *Right,* a narrow beam of light causes those brittle stars which cross it to huddle together. In their native habitat, brittle stars in bright light cling closely to the eel grass which matches their greenish color, but wander away during darkness. The "thigmotropism" is actually part of a complicated adaptive behavior pattern. (Original)

tropism," or a tendency to touch things. When we put the tank in a dark room and suddenly turn on the light after several hours, we find that they have left the corners and are scattered all over the bottom. Their behavior now begins to be understandable. These particular brittle stars normally live in eel grass in shallow water, and their long greenish arms are very much like the grass they live in. When exposed to bright light in the laboratory, they behave as they would among eel grass. They get into contact with as much solid surface as possible and lie quietly, a response which probably helps to save them from predators.

Reflexes. Even in the higher animals certain bits of mechanical behavior can be isolated. Such are the erection of the hair in response to cold and the salivary reflex which Pavlov studied in connection with the problem of learning. In many cases the adaptive value of these reflexes is quite obvious; in others they at first appear to be incomprehensible. A good example is the knee-jerk reflex in human beings, which can be elicited by a blow on the tendon just below the kneecap. At first glance it appears to be an entirely useless bit of behavior. However, if one stumbles on a rock, he can feel this reflex immediately coming into action and serving a useful purpose. Catching the foot puts a strain on the tendon, and the reflex produces a very quick straightening of the foot which is often sufficient to prevent a fall. Like tropisms, reflexes can be understood best as parts of larger patterns of behavior.

Patterns of behavior. Simple reflexes or tropisms often appear to have simple and obvious adaptive value as parts of general patterns of behavior. However, the significance of the behavior patterns themselves may not be immediately apparent, as the example of bird song shows. Nikko Tinbergen points out that the adaptive nature of behavior can be quantitatively measured under field conditions. For example, black-headed gulls, like many other similar sea and shore birds, nest in dense colonies in which a bird in one nest can almost touch that in another. In the colonies which Tinbergen and his co-workers studied along the coast of England, the nests are often attacked by herring gulls and carrion crows that eat the eggs. When a predatory bird approaches their colony, the black-headed gulls attack it as a group, and they are more successful in spots where large numbers of parent birds are available to repulse the intruder. The result in one study was that 291 eggs were stolen from the outside half of the colony while only 95 eggs were stolen from the inside half. Obviously those birds that make their nests close to those of others have a better chance to protect their eggs and raise offspring successfully.

Even the possible adaptive value of a behavior pattern is not always immediately obvious to the observer, particularly if it occurs in an artificial laboratory environment. In one of Tinbergen's laboratories, he accidentally set up a row of aquariums containing male three-spined sticklebacks in front of large windows overlooking a main street in the city of Leiden. The fish would sometimes dash to the window side of the tanks and attempt to swim through the glass. It turned out that they only did this when a red mail truck drove past. Eventually Tinbergen discovered that the red underbelly of a male stickleback acts as a signal which induces fighting in other males. In the original laboratory situation, the behavior had no function, and its significance could only be determined by a long series of experiments. Thus the problem of determining the adaptive function of a particular behavior pattern is dependent upon a second question: To what is the animal responding? Elaborate experiments are often necessary before any conclusion can be drawn, as in the examples given below.

THE NATURE OF EXTERNAL STIMULATION

The English ornithologist David Lack has described a series of experiments on the nature of the stimulus which causes the English robin to fight. It should be remembered that the English robin is quite a different bird from the American one which goes by the same name. The latter is a large thrush of relatively peaceable habits, often seen pulling worms out of lawns. The English robin redbreast is a more pugnacious and brightly colored member of the thrush family and a much smaller bird. When English visitors first see the American robin they are often led to remark, "Just like the Americans; everything twice as big as it is in England!"

Lack bought a stuffed robin and mounted it on a branch near the wild birds. During the winter they paid no attention to it. In the spring he put it next to pairs of birds which were building their nests. They began to attack it, and three females in succession deserted their nests as a result. The stuffed robin stimulated fighting even more successfully at the time when the young nestlings were just beginning to grow their feathers. The parent birds would leave the nest to attack the model but did not desert their young.

When the stuffed bird was left near the nest, the parents attacked it fiercely at first, but after several days they came to pay no attention to it. Here again we see the phenomenon of habituation. Even when the model was repeatedly taken away and returned, the responses became weaker and weaker.

Lack then tried to find out what it was about the stuffed robin that stimulated fighting. He began taking parts of it away and finally ended up with a bundle of red feathers above and white feathers below. This was attacked by half of the birds to which it was shown and ignored by the rest.

The results can be interpreted in a variety of ways. The simplest is to assume that we are here dealing with a simple mechanical bit of behavior similar to a reflex or a tropism, and that any robin which sees a bundle of red feathers in the breeding season automatically has to fight. But what about the 50 percent which were not so stimulated? One could get approximately the same result by giving a punching bag to a group of small boys, but we know that theirs is not simple reflex behavior.

Indeed, the process of learning is obviously present in Lack's experiment, and it raises certain problems which can be decided only by further experiments. The reaction to red feathers may be a simple primary reaction which later becomes modified by learning, just as reflexes can be associated with new stimuli. It is also possible that the situation is more complex, and that the birds at first react to any other fighting robin and only later come to associate fighting with certain prominent features of their opponents.

Fig. 20. An English robin threatens part of a stuffed model put near its nest. Only about half the birds will attack this bundle of feathers, but most birds will attack a complete stuffed model. (From a photograph by H. M. Southern, in David Lack, *The Life of the Robin*, by permission of H. F. and G. Witherby Ltd., London)

Choosing between these two explanations can be done only through an experiment using animals whose previous history is known, a thing which is rarely possible in field experiments or with animals raised in the wild.

Primary stimuli or releasers. Shortly after a baby chick emerges from the shell, it begins to peck at various objects. Some of these may be grains of food, but the chick may also peck at grains of sand and the eyes of other chicks. It is obvious that this reaction can be elicited by a rather large variety of objects which have in common only their relatively small size. A few days later the chick no longer pecks indiscriminately at inedible objects but prefers food, and as it grows older it can be taught to distinguish between colored grains of corn, one color being glued to the floor and the other being loose and easily eaten. This kind of behavior seems to be typical of the young of higher animals which start life with a few simple primary behavioral reactions. These can later be modified by experience and learning and organized into more complex patterns of behavior.

In his book on instinct, Tinbergen describes the intricate and intriguing process of finding out exactly what stimulates these primary reactions. A newly hatched thrush can be stimulated to open its beak by various means. In the natural situation the mother arrives on the nest, the birds open their beaks, and she sticks food inside. Before their eyes are open, the young thrushes react as soon as the nest shakes. Later, when their eyes open, the sight of a moving object nearby will produce the same effect. A stick or the human finger will work as well as the parent bird, but it must move, have some appreciable thickness, and be above the eye level of the nestlings. Anything vaguely approximating the movement of the mother bird's head will act as a sufficient stimulus.

A much more specific stimulus is found in young herring gulls. When the mother returns to the nest and puts her head down toward the young, they peck at her bill in a food-begging reaction. Tinbergen tested the young gulls with model heads. Normally there is a red spot at the tip of the lower bill. When he transferred the spot to the top of the head, the young chicks gave the food-begging reaction only one-fourth as often. From this and other experiments Tinbergen concluded that the bright color markings peculiar to certain species of birds are specific primary stimuli which aid in coordinating their social behavior.

This brings up the question of how we can identify a primary reaction as opposed to one which has been affected by previous learning. There are many instances of complicated patterns of adaptive behavior exhibited by adult animals the first time they are exposed to a particular stimulus. However, in any vertebrate animal, a primary response can be affected by the various processes of learning as soon as it appears, and in the case of any adult animal there is always the possibility that the reaction has been affected by earlier learning and experience. Even if the pattern as a whole has not appeared, it frequently turns out that parts of it have been used in earlier life. This is one of the main points of disagreement regarding the effect of heredity upon behavior. It is conceded by everyone that a young animal can be stimulated to simple behavior without training or experience, but can an adult be similarly stimulated to produce highly organized patterns of behavior?

The answer probably depends in part upon the kind of animal which is being studied. For example, it is often said that sheep dogs have an instinct to herd sheep, meaning that they will begin to herd sheep, or even other animals, with little or no experience. However, in practical sheep management the dogs are subjected to long periods of training, being given not only contact with the shepherd and the sheep but associating with older dogs which have already been trained to herd.

In our laboratory we tried to teach one of our Shetland sheep dogs to herd a flock of goats into the barn. This particular individual was several years old and had always lived with dogs and people, without direct contact with other animals. His reaction to the goats was to pay no attention to them whatever. At first the goats were afraid of him and ran ahead, but he still made no response to them. He eventually worked out a pattern of behavior consisting of running into the barn ahead of the goats every time any attempt was made to train him to herd. The goats, of course, soon learned to run in the opposite direction.

One explanation of his behavior might be that we had accidentally picked an animal with poor hereditary abilities, but Dr. Ginsburg then took other young sheep dogs and tried them out on the goats at six months of age. Their reaction was to run after the flock and bark in a way which looked like rudimentary herding. Other breeds of dogs did the same thing, the terriers biting the goats as well. This chasing provides a starting point for the process of training the animals to work under direction, but it is still a far cry from the complex activity of a trained sheep dog.

We put the sheep dogs through our regular testing program, from which it appeared that these animals have a number of characteristics which make it easy to train them to herd sheep. One of these is aggressiveness, but at the same time the dogs must be sensitive to noises, so that they will pay attention to commands and can be easily inhibited. They have strong emotional reactions and form close associations with human beings. All of these capacities add up to an animal which can readily organize its behavior into herding, but there is no evidence that this complicated pattern of behavior is inherited in a preorganized unit. As the behavior of the older dog showed, this behavior can be organized in other ways which actually interfere with learning to herd.

We might also think that the tendency of a cocker spaniel to crouch or sit and stay quiet is such an inherently organized piece of behavior. However, our cross-breeding experiments showed that the sitting posture and the tendency to remain quiet were inherited in different ways. Even this apparently simple behavior trait is actually a compound one. Like the sheep dog, what the cocker actually inherits is the ability to develop a combined pattern from several traits, and this makes it easy to train him to sit quietly when he has flushed birds and his master is shooting at them. The process of organization is partly dependent upon these traits and partly dependent upon the kind of problems of environmental adaptation which confront the animal. The dog trainer can make it easy for the animal to learn, but it is the dog itself which does the learning and so organizes its behavior.

The sexual behavior of a dog might also be thought of as a complicated behavior pattern released by a primary stimulus late in life. However, it is observed very early in their development that young puppies roll over and over each other, clasping and mounting their playmates in ways which duplicate parts of the pattern of sexual behavior, although the pattern itself is not always recognizable. In our laboratory, A. A. Pawlowski raised a pair of female beagles in separate cages so that they never came into contact with each other or with other dogs until they became mature. He then proceeded to test the theory of primary stimulation by exposing them to male dogs at the time of the first estrus. As the males attempted to mount them, the females reacted with an inappropriate response, rolling over on their backs and waving their feet in the air. It was only after long contact with experienced males, who more or less held the females in the proper position, that mating was achieved.

Later Frank Beach did a much more extensive experiment with beagle males reared apart from other dogs. On becoming adults these males were often ineffective in their mating behavior, mounting females from the wrong end and never achieving complete copulation.

Among primates, chimpanzees reared by hand, as was formerly a regular practice at the Yerkes Primate Laboratory, were unable to mate as adults except with animals having had previous experience. An even more drastic effect was discovered by Harry Harlow and his associates when they raised rhesus monkeys in isolation. In this species copulation is possible only with the male standing on the female's ankles, so that coordination is necessary in both partners. Even experienced animals were for the most part unable to induce coordinated sexual behavior in the inexperienced partner. On the other hand, monkeys that had had the experience of playing with other young monkeys for a few weeks were able to function normally as adults.

When mating was finally achieved with some of the isolated females, they showed gross defects in maternal behavior after their young were born. Instead of constantly holding an infant in her arms, a mother reared as an isolate would neglect it, step on it, or even strike it when it called. However, maternal behavior improved considerably after a second pregnancy. Thus there is considerable evidence that the organization of sexual and maternal behavior in some of the higher mammals is dependent upon experience gained after birth, and also that coordination and efficiency of these behaviors improves with adult experience.

The apparent contradiction between results with mammals and those obtained with birds may be simply a matter of degree of genetic control. Birds do seem to exhibit complex patterns of coordinated behavior as primary reactions. The behavior of birds is in many ways quite stereotyped, which goes along with relatively rapid development preceding life in an aerial habitat and a relatively poor development of the neocortex of the brain. In contrast, mammals seem to give responses to primary stimuli only in early development and usually with quite simple forms of behavior which are later combined into elaborate patterns in a variety of ways. The real scientific problem is how far behavior in any particular species can be organized by psychological means as opposed to organization on a physiological and genetic basis.

The animal's sensory world. Apart from this basic scientific question, the process of experimenting with the sources of primary stimulation gives us a picture of the parts of the environment to which an animal normally responds. At the very least we may conclude that different species respond to different features in the environment.

The American psychologist Lashley, in his early experiments on problem-solving in rats, used an apparatus in which there was a choice between two alleys. As the rat was put into the apparatus it could see both alleys, each of which was marked with a different sort of symbol. The rats did not learn to distinguish between them, and Lashley came to the conclusion that the rats either had defective eyes or were not using them. It seemed rather improbable that an animal with functional eyes should fail to use them at all, and after various kinds of experimentation Lashley finally came up with an apparatus in which rats could solve a problem only by the use of their eyes. Each rat was put on the top of a pedestal with two doors in front of it. At first the pedestal was placed close to the open doors, through which the rats could walk and obtain a bit of food. The next step was to close the doors but fix them so that the rats could push them open. Then the pedestal was moved away, making it necessary for the rats to jump to get through the doors, knocking them down in the process. Now the doors were marked with different symbols, such as squares or triangles. If the rats jumped to the correct one, it opened and they got food, but if they made the wrong choice, they simply struck their noses against it and fell into a basket below. Under these circumstances the rats quickly learned to recognize the sign for the correct door. This demonstrated that the rats ordinarily used their senses of touch and smell to solve a problem and did not bother to use their eyes unless it was absolutely necessary. Since the rat is largely nocturnal in its habits, this is probably the way in which most of its behavior is organized.

This illustrates a further point about interpreting animal behavior. In spite of having a variety of stimuli presented to it, an animal may respond to only one or a small part of the total situation. The fact that it does not respond to a particular stimulus under a special experimental situation does not prove that it cannot. We should neither exaggerate nor underestimate the powers which an animal has for dealing with its environment.

The many studies of bird behavior lead us to conclude that the world of a bird is predominantly a visual one, as we might expect from the fact that the majority of them are active during the day-

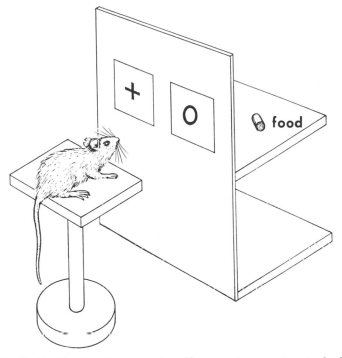

Fig. 21. Lashley jumping apparatus. There are two openings in the board in front of the rat. At first its platform is pushed close so that it can walk through and get food. Next it learns to push a card aside to get through. Then the platform is moved back so that the rat has to jump. One of the cards, marked "+," falls down easily, but if the rat hits the other, the animal falls to the table. Under these circumstances the rat uses its eyes to discriminate between the two cards. (Original)

time and have highly developed eyes. Tinbergen gives a delightful picture of the sorts of things which make up the world of a herring gull. The bright colors and conspicuous markings of these and many other birds are related to their predominant response to visual stimuli. The ears of birds are not so obvious, but auditory stimuli run a close second. A feeding sparrow takes flight at the slightest sound, and geese on the wing call to each other continuously.

The range of variability in the use of sense organs is much greater among mammals. In many cases there is a great variety of sensory reactions in the same species. A herd animal like the deer has highly developed eyes, but it also uses scent and sound, as any hunter knows who has tried to stalk one from an upwind direction. Dogs have the reputation of being animals which make predominant use of their noses, but they also have excellent ears and make at least moderate use of their eyes. Man and his primate relatives tend to specialize in vision and to use hearing and smell in a secondary way. In studying the behavior of other animals, man is often led astray by his own visual bias.

When we attempt to explore the sensory world of an invertebrate, we get onto more difficult ground, since their sense organs are so different from those of human beings and other vertebrates. Indeed, for many years the use of certain organs of insects could not be understood at all. Some success has now been achieved through modern physical and experimental methods. It is possible to take photographs through the eye of an insect and thus find out the range of focus and kind of picture presented. It is also possible to record sounds even if they are inaudible to the human ear and play them back to the insect. All of us are familiar with the unpleasant whine which a female mosquito makes when she is about to alight and insert her proboscis. Another sound serves as a mating call, and when it is played back to other mosquitoes, the males fly toward it. It has been used with some success in the construction of mosquito traps, luring mosquitoes not to a mate but to an electrified grid which quickly disposes of them. The practical value of the trap is dubious, since it gets only the males. Male mosquitoes do not feed on blood and so do not annoy us anyway, and a single male which escapes can fertilize several females.

When we consider all these points, we see that each species is selective in the stimuli to which it responds in the environment. This is partly caused by the nature of the sense organs but is also related to the type of life which the animal leads. The senses most employed are those which are most useful in the particular environment. As we discover this information, we find that an animal is not a small furred or feathered human being but another kind of individual which lives in a world of its own, of which we can sometimes get a glimpse by experimental means.

THE ORGANIZATION OF BEHAVIOR

The third question which was raised at the beginning of this chapter is: How does an animal organize its behavior to meet a new problem of adaptation? As indicated above, its behavior may already be partly organized in early development, under the influence of heredity and the prenatal or prehatching environment, so that an animal comes into the world with standard ways of meeting certain common difficulties. On the other hand, animals frequently organize their behavior in unique ways and seem to show what is called in common terms real intelligence.

The story of Clever Hans. About the year 1900 stories began to circulate in Germany about an extraordinarily clever horse named Hans. He would stand in front of his trainer and paw the ground with his hoof, and if you asked him the sum of two and two, he would paw the ground four times. He could not only do addition but also multiplication and division; he could spell out words and sentences by pawing the ground the appropriate number of times for each letter of the alphabet. Remembering that R is the eighteenth letter of the alphabet in the midst of spelling a complicated sentence is a pretty difficult task, and Hans was all the more remarkable because he learned all this in the space of two years. It was possible to conclude that Hans was not only a very clever horse but that he was a good deal smarter than many children.

A committee of zoologists and psychologists studied Hans and found that the horse could indeed do the sort of thing which had been reported. One of the first hints of how Hans got his results came when they found that he always failed when no one present knew the answer to a problem. This suggested that the master, who apparently was standing perfectly still and waiting for the answer, was in some way giving Hans an unconscious signal when he got the right answer. Sure enough, when a screen was put between master and horse, Hans lost his powers entirely. All that really happened in the case of this wonder horse was that he had been taught to paw the ground, and if he pawed long enough, he would inevitably come to the right answer. At this point his master would feel relieved and relax slightly, and Hans saw that this was the time to stop. Hans was a highly trainable and observant horse, but he could not do arithmetic.

This illustrates the importance, in any experiment involving animal intelligence, of having the experimenter remove himself from the situation so that he cannot give the animal unconscious encouragement and help. The best experiments, of course, are done in such a way that the experimenter himself does not know the right answer. When these precautions are taken, the facts are interesting enough without assuming any miraculous powers of animals.

Organization through trial and error and habit formation. As we saw in an earlier chapter, a great deal of adaptive behavior can be explained on the basis of variability of behavior, accidental solution of a problem, and subsequent habit formation. If a puppy is given the problem of getting food from a dish which is covered up, it tries all sorts of things: sniffing at the cover, standing on it, pawing it, biting it, and trying to lift the cover with its nose. Eventually the puppy tries something which works. On succeeding trials it has two tendencies, one to go through the whole performance again, and the other to vary the behavior by trying things in a different order. In this way it gradually eliminates the things which did not work until it finally does the task very efficiently. By varying its behavior, the puppy is able to discriminate between those parts of the job which bring success and those which do not.

Solving problems without trial and error. In the year 1913 a young German psychologist named Wolfgang Koehler had the chance to go to the island of Tenerife in the Canaries and study the behavior of a colony of anthropoid apes maintained by the Prussian Academy of Science. Interested by the fact that adult animals frequently solve a problem the first time they meet it, he devised a kind of experimental technique which may be called the "detour problem." A chimpanzee is shown a banana, but there are obstacles in the way. Instead of trying out various possible paths, the animal immediately takes an indirect path which allows it to get to the goal quickly. The behavior of the animal is apparently organized before it does anything, and Koehler explained this as a *Gestalt*, which has been translated as "configuration." What this means will perhaps be made clear by an actual example.

In our laboratory we tried a very simple variation of the detour experiment on six-week-old puppies. They had been raised in a rectangular room containing no barriers with which they could have had previous experience. They were taken to a different room and fed from a small dish several times on two successive days. On the third day a high fence six feet long was placed in front of the dish and a puppy put on the opposite side. The puppy

Fig. 22. Reactions of young puppies to a detour problem. *Above,* puppy at first tries to get at the food directly, scratching at the wire and vocalizing, but eventually finds it way around the end of the barrier to the food. *Below,* two solutions to a more difficult problem. The barrier has been extended. One puppy, having learned the solution by distance, attempts to go through the place where the barrier formerly ended. Another, having learned a visual rather than a spatial solution immediately goes around the end.

could see the dish through a hole in the fence as long as he stayed near it, but the rest of the fence was opaque. Under these circumstances the puppies showed a variety of reactions. In the most successful cases they made a brief effort to get through the fence, then walked to one end and looked around, then came back to the other end, looked around it, and finally went to the dish. Their behavior suggested that they were investigating the situation and finally arriving at a solution after investigation.

Other puppies were not so successful, and many of them would make a few short trips away from the dish and then become highly excited, yelping and dashing back and forth ineffectually. As long as they were excited and yelping, their behavior was variable but stereotyped, and they were never able to solve the problem. This suggests that such behavior may be a primary adaptive response which is largely governed by heredity. Finally they would stop yelping, look around quietly, and then often make a successful exploratory trip down to the end of the barrier.

After they had once succeeded, the puppies were given two more trials to get used to the situation, and they usually showed considerable improvement. On the next day the problem was made more difficult by adding two more sections of fence on the ends so that it was now eighteen feet long. Some of the puppies dashed around the end immediately, which looks very much like the findings of Koehler where there was immediate solution without trial and error. Other puppies would simply run as far as the end of the original fence, be unable to get through at this point, and then become very much excited and revert to yelping and dashing back and forth. The simplest interpretation of the behavior is that one set of puppies had associated success with running to the end of the fence wherever it might be, whereas the others had made an association of running only a short distance. One of these happened to be the solution for the next problem and the other was not. In either case it looks as if the puppies had organized their motor behavior so that their first reaction to a new situation was not one of trial and error.

Koehler did a great deal of his work with chimpanzees, which have much greater manipulative skill than dogs. He found that they were able to solve problems which were highly complex and almost impossible to solve by simple trial and error. In one case he hung some food high out of reach and placed a number of boxes in the cage. The chimpanzees were able to stack these on top of each other and reach the bananas. In other experiments Koehler placed bananas outside the cage and gave the chimpanzees a stick with which to rake in the food. In some cases they were even able to take two sticks that fitted together and make a longer one to reach their reward.

These and other experiments with problem-solving suggest that animals are capable of a considerable degree of organization of behavior, apart from that set up during the process of embryonic development and therefore largely governed by heredity. One of the simplest kinds of organization, and one preferred by most animals, is that of motor behavior or organization on what might be called a kinetic basis. A mouse which has thoroughly investigated a pen with a series of complicated passages will dash through them at top speed when pursued by another mouse and blindly run into new obstacles which are placed in its path even if these are perfectly visible. The mouse has organized its behavior into a memorized series of movements.

However, the experiments of Koehler suggest that some organization of behavior is possible on other bases as well. When confronted with a new situation, chimpanzees are apparently able to

Fig. 23. Superior manipulative capacities enable chimpanzees to solve problems requiring the use of tools. *Left,* Jojo fits a short pointed stick into the hollow end of a longer stick. *Right,* she uses this tool as a spear for reaching through the wire to get food which she cannot get with her arm. (From photographs of a chimpanzee at the Yerkes Laboratories of Primate Biology by Catherine H. Nissen)

relate it to their previous behavior. They work out a solution in a way which might be similar to what we experience as visualization, in which we see an imaginary picture of things as they might exist. The chimpanzee who reaches with a stick may see the stick as a long arm, or perhaps feel it as one. Once this has been done, the solution is achieved almost instantly.

Tool using in animals. Tool using in chimpanzees was originally thought of as a product of being placed in an artificial environment, that is, as behavior that was essentially forced upon the animal. However, chimpanzees living under natural conditions also use tools. Jane Goodall has observed chimpanzees breaking off twigs, sticking them down holes in termite nests, then withdrawing the twigs and licking off the termites that cling to them. Chimpanzees will also pick up sticks or branches and throw them at threatening predators. Adriaan Kortlandt put a tame leopard on top of a wall of an enclosure in which several chimpanzees were kept under semiwild conditions. The adult chimpanzees immediately responded by screaming and yelling, jumping and stamping with their hands and feet and then by throwing sticks at the leopard. Later Kortlandt placed a stuffed leopard, apparently holding a baby chimpanzee, under a bush in an area where wild chimpanzees were living. When a band appeared he hauled the leopard into view and the chimpanzees began making charges, sometimes bare-handed but often with sticks which they brandished or threw at the stuffed leopard, but never coming very close to it themselves. There is no doubt that chimpanzees use objects as tools under a variety of situations, and this gives us a model of how our own early ancestors may have acted.

Primates are not the only animals that use tools. One of the solitary wasps (*Ammophila urnaria*) uses a small pebble as a hammer to pound dirt into the nest burrow. One of the most remarkable instances of tool using in an invertebrate is a spider which builds an elastic web, then draws the center of it back in a cone shape. When a flying insect appears the spider lets go the net, which springs out and catches the insect. However, in both these cases there is every indication that tool using is simply an extension of the regular patterns of behavior of the species rather than arising as a solution to a problem.

Among vertebrates many animals other than primates use tools.

Fig. 24. Tool using in California sea otters. *Left,* otter eats an abalone from its single shell. *Right,* another otter hammers a clam shell on a stone which it has brought up from the bottom. The stone is a primitive tool.

One of the most remarkable is the woodpecker finch of the Galapagos islands. Like a woodpecker, this bird climbs up and down trees looking for food. When it comes to a hole which may contain an insect it picks up a cactus spine or twig and pokes it into the crack, levering it around until the insect larva emerges, whereupon the bird drops the twig and eats the larva.

Still another example is the sea otter that lives off the western coast of North America and dives for shellfish such as abalones, mussels, clams, and crabs. The abalone, which has only one shell, is levered off a rock and brought to the surface, where it is eaten, the otter lying on its back in the water and holding up the shell very much like a person licking off a dinner plate. When it brings up a mussel it may also bring up a flat rock, place the rock on its chest and use it as an anvil on which to break the mussel shell. It holds the mussel in both hands and brings it down hard on the rock. A sea otter can also crack clams or spiny lobsters in the same way. Since the young sea otters stay close to the mother for many weeks, clambering on her chest and taking food from her as well as nursing, they have ample opportunity to learn the use of the tool and to discriminate between those shellfish which require cracking and those which do not. However, no experiments have yet been done to determine whether the tool using is part of the regular behavioral repertory of the species or whether it is passed along in the manner of cultural inheritance.

Testing "intelligence" in animals. In the broadest sense, the concept of intelligence may be defined as the organization of behavior. Consequently if one wishes to study intelligence, he must try to find out how behavior is organized in a particular animal species or group of species. The most general technique is to present the animal with a situation that requires adaptation and see how he performs. This can be done under natural conditions, as Kenneth Gordon did with chipmunks. These small rodents are relatively tame and hoard food. Presented with a problem of obtaining food indirectly, such as by string pulling, the chipmunk will literally work all day for peanuts, since he never eats them, but simply carries them away, stores them, and comes back for more. Or the testing can be done under the highly artificial conditions of the laboratory, which permit much better control over environmental conditions but also present problems of emotional adjustment. For example, a favorite piece of testing apparatus for primates and other animals capable of manipulation is one in which the animal sits in a cage behind bars and can reach out to seize one of several objects that may conceal food. Thus a monkey can learn to find food when the only cue given him is that one of three objects is distinctly different from the other two. Or there may be three square pieces of wood, with the food always concealed beneath the middle-sized one. However, getting wild monkeys used to the apparatus and interested in the problem may take weeks of effort.

The kinds of apparatus that can be used are only limited by the ingenuity of the experimenter. For example, the Russian animal behaviorist L. V. Krushinski set up a situation in which food is carried on a toy mechanical train that runs through a tunnel. The animals being tested see the food disappearing into the tunnel, and some of them are capable of immediately going to the other end and waiting for the food to come out.

In order to give meaningful information about the organization of behavior, the following conditions must be met. The problem must be one which is suited to the sensory and motor capacities of the individual. Manipulation tests are appropriate for primates and raccoons, but highly inappropriate for hoofed animals like sheep. Second, the animal must be motivated in some way to perform the task. Third, as we have seen above, emotional reactions, particularly those of fear, can easily disrupt adaptive behavior. The animal must be confident and secure in the situation. Finally, many species are capable of solving highly complex problems by the use of stereotyped behavior patterns. The way in which such complex behavior is organized can only be determined by a series of experiments in which heredity and previous experience are systematically varied and controlled. The essential principle, as stated in the previous chapter, is that behavior is developed under the influence of a variety of factors, and as it develops it becomes organized through a variety of processes.

One result of these considerations is that it becomes very difficult to rank different species of animals in terms of pure intelligence—meaning capacities of the central nervous system. One can rank them on their abilities to accept certain kinds of training and on their performance of certain tasks, but the results may depend as much upon sensory and motor capacities, emotional and motivational reactions, and developmental history, as upon basic differences in the capacities of their central nervous systems.

There are various ways in which behavior can be organized by an animal. These can be verified experimentally only in an indirect way, and great precautions need to be taken to avoid the mistakes of anthropomorphism and unconscious human help (the Clever Hans error). At any rate, it can be concluded that animals, and particularly mammals, are capable of a considerable degree of organization of behavior without the use of verbal symbols such as are used in human reasoning. Indeed, many results suggest that the use of "reason" or verbal thinking in human beings is not as frequent as is commonly supposed and that in many practical situations people may make use of the more primitive kinds of behavioral organization found in animals. Verbal reasoning is often a slow and cumbersome process and unsuited for use in practical emergencies. The usefulness of language lies primarily in communication, and the importance of a verbal solution for many problems is chiefly that it can easily be passed along to others.

The effects of early experience. Observing rats as they lived in small barren cages in his laboratory at McGill University, Donald Hebb one day developed the idea that they might perform better on tests involving complex behavior if they had a richer environment. He therefore took several young rats home and had his children bring them up as pets, exposing them to a variety of conditions and experiences. Such rats performed much better on the Hebb-Williams maze than those that had been reared in the laboratory cages. Later, Hebb's associates, raising rats in large cages furnished with a variety of objects which the rats could climb, manipulate, and explore, showed that there was a critical period, shortly after weaning, in which exposure to an enriched environment produced a maximum effect.

Watching our dogs being reared in the "school for dogs" experiment described in the last chapter, Hebb thought that even they were being reared in a relatively barren environment, and he decided to test the effects of an extreme experiment in which Scottish terrier puppies were reared in isolation beginning in early puppyhood and for many months thereafter. When they emerged, the terriers were extremely odd animals, learning very poorly in most situations and even being unresponsive to painful stimulation such as hot steam radiator pipes, which an ordinary dog would instantly avoid.

Later John L. Fuller and Lincoln Clark repeated Hebb's isolation experiment on a shorter time scale, from three until sixteen weeks of age. These puppies also were deficient in learning capacities, but it was even more obvious that they were badly disturbed emotionally by emerging from the safe and familiar boxes in which they had spent all their early lives. These isolated animals had no previous experience with either separation from familiar objects or contact with strange situations, either of which produces strong emotional distress reactions in a normally reared dog. Thus the effect of removal from a familiar environment was to give the puppy its first experience with a strongly unpleasant emotional reaction which thus became associated with the entire world outside the familiar cage in which it lived. The implication is that the defects in learning capacity observed in these pups reared in barren environments arose as much from emotional reactions to sudden change as from a lack of previous learning experience that could be transferred to new problem solving.

Such deficiencies in early experience have an even more profound effect upon subsequent social behavior and the organization of social relationships, as will be described in the next chapter.

CONCLUSION

So far in this book we have pursued the idea that behavior has certain causes. We have systematically examined these and can now come to the general conclusion that the causes of behavior are found at any level of organization and that factors on each level are interconnected with those on other levels.

Beginning on the individual level of organization with behavior itself, we find that behavior is essentially adaptive. Examination of the behavior of all sorts of animals shows that there are a limited number of basic systems of adaptation such as ingestive behavior, sexual behavior, and the like, each of them with causes or stimuli. The most important general theory of behavior is the stimulus-response theory. In the most general sense a stimulus is a change and a response is an attempt to adapt to change. The most important cause of behavior is therefore an environmental change.

THE ORGANIZATION OF BEHAVIOR

LEVEL OF ORGANIZATION	UNIT	EFFECT ON BEHAVIOR
Ecological	Population	Localization, territoriality, etc.
Social	Society	Socialization, dominance, leadership, etc.
Organismic (Psychological)	Individual	Behavioral adaptation, learning, psychological organization, and intelligence
Physiological	Organ System	Internal changes, physiological adaptation
	Organ	Sensory and motor capacities
	Cell	Transmission of stimuli; motion
Genetic	Gene	Primary stimuli and responses, trait complexes

When we follow the causes of behavior into the level of organs and organ systems, we find that sensory and motor capacities have great effects upon behavior. Of particular interest is the ability to grasp and manipulate objects. An animal which can do this can do more than adapt itself to the environment; it can change and adapt the environment to itself. Consequently, animals with well-developed manipulative abilities have a wide range of adaptation and a good reputation for intelligence.

On the same level of organization, we can see that behind every major system of behavior there is a long chain of physiological causes. In ingestive behavior there are internal changes which arise simply from the processes of living. Such stimuli as hunger are relatively independent of external changes and produce important effects on behavior. In other types of adaptation, such as agonistic behavior, internal changes are almost entirely controlled by external stimulation.

Studying the effects of such changes, we find that the behavior of almost all animals is affected by previous as well as immediate stimulation. The ways in which behavior is modified by previous experience can be expressed in a few general principles. The first of these is the variability of behavior in response to repeated stimulation. Some reactions produce successful adaptation and some do not. An animal associates the failure or success of its attempts at adaptation with its response to previous stimulation, so that behavior tends to become more and more highly adaptive. Habit formation decreases the variability of behavior, so that the behavior of animals capable of a high degree of learning can become constant and predictable.

The sensory and motor capacities in different species are determined by growth and development, which in turn are strongly influenced by heredity. There are other more direct ways in which heredity can produce effects on behavior. We found evidence that heredity can produce important effects on thresholds of stimulation and response. These in turn strongly affect the motivation of an animal and its ultimate capacity for adaptation. The effects of heredity can eventually be traced back to the genes, which affect behavior on the most basic level of organization.

In this latest chapter we have made a full circle by coming back to the level of organization of individual behavior and the problem of how heredity can affect it. Many animals are capable of a great degree of psychological organization of behavior, as when an animal solves a problem in a new way. This ability itself differs from species to species and from group to group, and we may conclude that it is strongly influenced by heredity. On the other hand, the primary action of the genes is often quite remote from behavior, and heredity must then produce its effect through long chains of physiological processes.

As we study the basic causes of behavior, we find repeated hints that behavior is also affected by causes on the higher levels of social and ecological organization. In order to understand them, behavior must be considered as a cause of the higher levels of organization as well as an effect. The preceding chapters are concerned with the causes of behavior; the following pages deal with its results.

CHAPTER 8

Social Behavior
and Social Organization

Chickens are one of the several highly social animals domesticated by man. Fifty years ago every farm family had a flock, and a chicken yard was a common sight in the back yards of small towns. The fussy behavior of hens and the early morning crowing of roosters became proverbial, but for the most part people took advantage of the social behavior of chickens without attempting to understand it. Then the Norwegian scientist Schjelderup-Ebbe observed that one hen in his flock was acting like a tyrant, always pecking the others and driving them out of the way.

He set out to study the social life of hens, and being a careful observer he was soon able to identify each one by its appearance and behavior. The hens did a great deal of pecking and threatening as they fed and walked around the yard, and he saw that in each conflict between two hens one always pecked and the other submitted without fighting back. The first hen was dominant over the second, and their behavior was called a dominance-subordination relationship. When he had described relationships between all the members of a flock, he saw that the whole group was organized into a "peck order" or dominance hierarchy.

Since then the observation has been repeated in many different situations with essentially the same results. For example, in a flock of only three hens there are three possible relationships, and the usual situation is one in which hen A attacks hens B and C, which do not fight back, and hen B habitually attacks hen C without retaliation. In a situation like this there is a straight-line dominance order from A to B and from B to C, and hen A is said to be dominant over the other two, which are subordinate.

Why did this social organization go unnoticed for so many years? As pointed out in an earlier chapter, an essential technique

174

in the study of animal behavior is the identification of individuals, and the peck order would never have been suspected if Schjelderup-Ebbe had not learned to tell the hens apart. All that the casual observer sees is a certain amount of antagonism between the members of the flock.

The origin of social organization of this type can be studied when two strange hens are put together for the first time. They usually fight vigorously, flapping and pecking. Soon one of the hens gives up and runs away. At the next encounter there is a short repetition of the fight, with the same animal tending to come out on top. In successive meetings there is less and less fighting until finally the dominant hen has only to threaten to peck the subordinate one to make it get out of the way.

At this point the two hens get along together with a minimum of fighting. A flock organized into a dominance order eats more and lays more eggs than a group of strangers. Thus a dominance

Fig. 25. Dominance relationships in a flock of three hens. *Above,* Alpha hen stands near food dish and drives away Beta hen with a peck, while Omega (bottom ranking hen) runs away. *Below,* as Alpha leaves the dish, Beta pecks Omega, who now tries to approach.

order may be thought of as an adaptation for reducing destructive fighting within a group.

One question immediately arises: Is a dominance order functional under natural conditions or is it entirely a product of crowding hens together under artificial conditions? The ideal place to study chicken behavior in the wild would be among the jungle fowl of Southeastern Asia, as these are the presumed ancestors of the domestic chicken. Much of the same behavior does go on in the jungles, as Nicholas Collias found, but the birds are scarce and very difficult to study in the dense vegetation. Glenn McBride discovered a colony of domestic chickens that had gone wild on a small island off the coast of Queensland in Australia. There was considerable vegetation but not enough to conceal the behavior of the animals. Furthermore, he was able to study the social organization of males, something which is quite difficult to do in the restricted space of a laboratory. During the breeding season in spring and summer, a small number of males maintained territories with fixed boundaries. Each male was attended by several females. In turn, each of their territories was divided into smaller territories held by subordinate males, each attended by females and each dominant in his own territory against lesser males. Still lower-ranking males might inhabit each territory, but these kept apart from the others and were not attended by females. Thus the females, under these seminatural conditions, are grouped into small flocks, each with a single male. Among themselves the females showed dominance at feeding sites but only at some distance from the male. All fighting among females was inhibited close to the male and this in itself may serve to produce group adhesion among them; that is, a hen close to a male does not get pecked. However, the fact that the hens did have a dominance order for feeding shows that the dominance orders seen in flocks of domestic fowls are not artifacts, and indeed, the older method of maintaining chickens, with one rooster and a flock of hens, is quite similar to organization under these seminatural conditions.

Since the publication of Schjelderup-Ebbe's work, dominance orders have been found in all the major classes of vertebrates and in many arthropods as well, and many of our present basic ideas about social organization have been developed from them. However, it should not be assumed that this is the only type of social

Dominance-subordination relationship in wolves. Animal on right is dominant, as shown by erect tail, that on left is subordinate, holding the tail down. (Photo courtesy of D. H. Pimlott)

Mother-young social relationship in a group of baboons. Several types of social behavior make up this relationship. Care-giving behavior is shown by grooming and suckling the young; these in turn show ingestive behavior and a social type of shelter-seeking. (Photographed in the Paris Zoo by R. Buchsbaum)

Nursing and the following reaction in a buffalo cow and her calf are part of the normal process of socialization in this species. Like most herd animals, she permits nursing only by her own calf, bringing about a close social relationship between mother and offspring. It also produces in the young animal a strong tendency to follow. In sheep this behavior has been shown to produce a definite system of leadership, and the same thing may also occur in wild herds. (Photo by J. A. King)

Care-dependency relationship. *Left:* normal rhesus monkey mother grooming her infant. *Right:* rejecting behavior by a "motherless mother"—one raised with only an inanimate model. (Photos courtesy Wisconsin Regional Primate Research Center)

Social behavior of the sage grouse. A strutting ground in early spring before the females appear. Several males are scattered around, exhibiting their remarkable display behavior. One of the two birds in the foreground is probably a master cock. (Photo by J. W. Scott)

Strutting or display behavior of the male sage grouse. During the mating season, these birds develop spectacular plumage and go through an elaborate pattern of behavior which displays the feathers to the maximum extent. (Photo by J. W. Scott)

Below, **a group of females gathered around master cock** and his attendant guard cocks. The master cock will do more than 80 per cent of the mating. (Photo by J. W. Scott)

Dr. Konrad Lorenz demonstrates the following reaction in a flock of greylag goslings. These birds can be socialized or "imprinted" so that they follow a human being rather than the natural mother. The experimenter must make contact with the young birds shortly after hatching and before they are allowed to see adult geese. (Thomas D. McAvoy—courtesy *Life Magazine* © 1955 Time Inc.)

organization or necessarily the most important one. There are a great many other ways in which the social behavior of animals can be organized.

THE DIFFERENTIATION OF SOCIAL BEHAVIOR

When we watch the development of social organization between any two hens, we see that what is really happening is a differentiation of behavior. When hens are first brought together, they all fight, their behavior is quite similar, and the flock is said to be unorganized. After the dominance order has been set up, certain hens peck and others submit to being pecked in a regular and predictable way, and the flock is said to be organized. The basic phenomenon of social organization is one in which the behavior between individuals becomes regularly and predictably different. Hen alpha always pecks, while hen omega is always subordinate.

Social behavior without differentiation. About the time that Schjelderup-Ebbe was making his studies of the peck order, a young American biologist named Allee was watching the behavior of a small crustacean, the isopod Asellus, as it swam in Indiana streams. Water isopods are poor swimmers, and when the spring floods come the currents are likely to carry them downstream. Under these conditions the isopods cluster together, hanging on to the bottom and to each other, and so are better able to resist the current. There is no evidence that particular isopods stick together; apparently any isopod will serve as a port in a storm. Every isopod behaves like every other, and the result is simply group formation without differentiation of behavior and hence without organization.

However, group formation is a necessary preliminary to social organization, and these temporary groups are the first step toward it. Allee went ahead to a lifetime of work in which he showed that physiological benefits can often result from the formation of simple aggregations. The type of adaptation involved is shelter-seeking, and it is probable that such aggregations, and similar groups formed by sexual behavior, were the original basis for the evolution of more complex social organization.

Simple unorganized groups also occur in animals which show a high degree of organization on other occasions. We remember that the redwinged blackbirds, which set up such a complex social organization during the mating and breeding season, also form immense autumn flocks which are relatively unorganized. Each bird simply reacts to the behavior of those next to it and all fly along together, so that the behavior of the whole mass is coordinated but undifferentiated.

The biological differentiation of behavior. An ant colony may be divided up into several classes of individuals based on differences in behavior. There are the winged males and females which originally start the colony and are the only ones which exhibit sex and reproductive behavior. Then there are the sterile females or workers which later take over the construction of the nest and the care of the young. Finally there are the young larvae themselves, which show only very immature forms of behavior.

In this example we can see three ways in which biological factors differentiate behavior. The first is hereditary differentiation of behavior based on sex determination. To this is added a differentiation based on feeding, which accounts for the sterile workers. Still a third type of differentiation of behavior is based on maturation, or growth and development of the young. Termites may carry this biological differentiation of behavior even further than the ants, with several classes of sexual individuals and with the workers divided into specialized classes of soldiers and nestbuilders.

By contrast the vast majority of animals, including vertebrates, are biologically differentiated into only three classes—males, females, and young—although it is possible to subdivide the young animals into groups depending on their stage of development.

The psychological differentiation of behavior. As far as biological differentiation goes, the members of a flock of hens of the same age are all alike, and their differentiation into dominant and subordinate animals is something which goes beyond biological determination. The differentiation of behavior which goes on when two hens come together is essentially a process of learning and habit formation and follows the general principles of learning described in a previous chapter. In the first stage there is considerable variability, and, if one hen attacks, the other can adjust to the situation either by fighting back or by running away if fighting fails. In subsequent meetings the adjustment of both birds tends to be reduced to a habit. Furthermore, the nonessential parts of the behavior patterns tend to be eliminated, so that in the end the dominant hen may make only a threat, and the subordinate one moves slightly out of the way. Since learning connected with fighting behavior has a tendency to be long-lasting, the dominance relationship is usually very stable.

This type of relationship can be contrasted with that in insects. For example, all worker ants tend to act very much alike in relation to each other. Even in ants, however, learning probably plays some part in the establishment of social organization, since if worker ants are carried away as young animals they will care for the young of the capturing species as devotedly as their own.

When the behavior between two individuals has become differentiated by either biological or psychological means and has become regular and predictable, we can say that a social relationship has been set up. Since there are so many possible ways of differentiation, it is obvious that many kinds of social relationships can exist.

Social behavior determines social relationships. Two things are necessary for the differentiation of behavior into a social relationship. One of these is the ability of an animal to discriminate between different individuals, and the other is some kind of behavior which can be differentiated by either biological or psychological factors. For example, a hen must be able to tell the other hens apart and must be capable of either fighting or escape behavior before a dominance order can be established.

We made a simple experimental test of the latter idea in another species by training mice to fight and not to fight. The fighting of mice can be inhibited by handling them just before they are put together, and we did this with several pairs of animals. They lived peacefully together for weeks thereafter. Then we took mice and trained them in the opposite way by having them first briefly attacked by other males and then repeatedly allowing them to attack helpless mice. This success-training made them into fierce fighters. When we put such trained fighters together, they at once had a fight, and one of the pair lost. Thereafter, whenever the pair met, the victor would chase the loser around the pen. When fighting was present, a dominance order based upon fighting was developed, and, when it was absent, there was no dominance order. It follows that social behavior is one important determinant of social organization and that the kinds of social relationships that any species develops will depend on the kinds of social behavior in its repertory. This point is well illustrated in the next section.

THE ORGANIZATION OF AN ANIMAL SOCIETY

When the Yerkes Laboratory of Primate Biology was founded in Orange Park, Florida, its directors were strongly interested in obtaining fundamental information about the behavior of as many primates as possible. They sent C. R. Carpenter, a young research fellow, to Barro Colorado Island in the Panama Canal Zone to study the wild population of howling monkeys. The result was one of the best general studies on social organization of higher animals that has ever been completed.

Carpenter first looked at the motor and sensory abilities of the howling monkeys. They live in groups in the treetops, and, as we might expect, their hands and feet and even their tails are developed for grasping. Their feet are almost like human hands, with an opposable big toe, but the hand itself is not well developed. When grasping a branch, the howler puts its thumb and index finger on one side and its other three fingers on the other. Their fingers are therefore somewhat clumsy in picking up fine objects. Like most other primates, the howlers use their eyes more than any other sense organ and are active chiefly during the day.

When Carpenter studied the daily cycle of behavior, he found that a group of howlers usually sleep all night in a tree. Early in the morning they get up, move around until they find a tree with food in it, and then settle down to feed for an hour or two on the fruit, twigs, and leaves. During the middle of the day they rest for a few hours, then move on and feed in another good place in the late afternoon. As darkness comes, they settle down for the night and stay until the next morning.

Howling monkeys show all the general systems of behavioral adaptation: ingestive and eliminative behavior, investigative, shelter-seeking, and all the rest. One of their outstanding peculiarities is the very small amount of fighting behavior which they exhibit. Where chimpanzees or rhesus monkeys would fight, the male howling monkey gives out a loud roaring cry, and the female makes a high-pitched yap like the bark of a fox terrier. They make

Fig. 26. Social behavior in howling monkeys. Two large males howl excitedly while a female rescues an infant that has fallen to the ground. The usual social group includes several adult males and females with young of assorted ages. (Sketch of behavior described by C. R. Carpenter)

a variety of other sounds as well. As social animals go, the howlers are a noisy lot. Allelomimetic behavior is very prominent since they constantly follow each other's movements while they wander about and feed, and the females show a great deal of care-giving behavior directed toward the young. These kinds of social behavior are organized into definite social relationships.

Female-young relationships. The most prominent relationship in the group is that between a female and her infant offspring.

She constantly responds to the young monkey, carrying it everywhere during the first year or so of life, picking it up if it falls on the ground, crouching over it at night to protect it from the cold or rain, and occasionally feeding it twigs as it grows older. Before it is weaned, at approximately two years of age, it often rides on her back with its tail coiled tightly around hers. The whole relationship may be described as one of *care-dependency*, consisting of a great deal of care-giving behavior on the part of the mother and, going along with this, the nursing and care-soliciting behavior of the young one. When the infant falls, it cries, and when it is restored to the mother's arms, it makes a sort of purring noise.

Female-female relationships. The adult females do not fight with one another and usually stay close together, forming a little group of females and offspring within the clan. When the clan moves, they tend to follow the males in a group, but there is no evidence of any specific leadership. Their relationship seems to be founded on allelomimetic behavior with little or no differentiation, all the animals behaving much alike toward one another.

Young-young relationships. Staying by their mothers in the group of females and offspring, the young monkeys have a good deal of contact with each other. They chase each other through the trees, sometimes biting and wrestling, and these playful contacts seem to be almost the only fighting that takes place in this species. Some sort of weak dominance relationships are developed, and the rest of the behavior is of the undifferentiated allelomimetic type.

Male-female relationships. The females have a true estrus period which lasts several days. When a female is in heat, she will approach any nearby male and initiate sexual behavior. The male stays with her until sexually satiated, and then the female moves along to another male. There is no evidence of sexual jealousy or that one male is preferred above another. The *sexual relationships* are therefore temporary and nonspecific. When the group moves as a whole, the males always take the lead, and a general *leader-follower relationship* exists between the two sexes.

Male-young relationships. The old males are usually indifferent to the young; but if a young monkey falls out of a tree, the males become much excited, and the whole group will howl vigorously until it is rescued. Presumably this has the effect of frightening predators. In addition, the males will sometimes pick up the young monkey if the mother is unable to do it. This behavior is evidence for a weak care-dependency relationship between males and young.

Male-male relationships. The males of any particular clan do not fight among themselves, but they roar at any other group or individual that comes near. They do this in unison, and we can see in this behavior a relationship of *mutual defense,* combining agonistic and allelomimetic behavior. The males of a group tend to stick together, and as they move through the trees each one will explore separately, looking for routes from one branch to another. When one male is successful, he gives a clucking noise, and the others come over and follow him, the slower females and their offspring bringing up the rear. There is no tendency for one male to take the lead more than the others, and this leader-follower relationship changes from tree to tree.

Specific and general relationships. An animal can form social relationships either with one specific individual or with a whole class of individuals. In the howling monkeys there is only one kind of relationship which is specific, that between the mother and her own offspring. The howlers are somewhat unusual in having so many general relationships which apply to a whole sex or age group. Even the sexual relationship appears to be a general one.

The basis of social organization. The howling monkeys have their behavior divided into three biological types, that of males, females, and young. This provides the clue to a systematic study of social organization: to take each type of animal and study its behavior in relation to every possible type, including its own. For a male there are three relationships: male-male, male-female, and male-young. The total number of biologically determined relationships is only six, since some of the combinations repeat.

The same sort of analysis can be applied to a more specific situation where we want to know the dominance relationships between all the members of a group. The total number of possible relationships can be stated as a mathematical formula which shows how the complexity of social organization is related to the size of the group. The simple arithmetical formula $n(n-1)/2$ for the total number of possible combinations in a group gives us the answer of three relationships for a group of three, six for a group of four, and ten for a group of five. Each additional member of a group adds a number of relationships equal to the previous size of the group. Thus an animal added to a group of 10 creates 10 new relationships. As the group becomes larger, the number of relationships becomes almost astronomically large; consequently most experimental studies are done with relatively small groups.

Going back to the general analysis according to biologically determined relationships, each of these can be subdivided depending on the number of kinds of social behavior exhibited. For example, the male and female howlers develop both a sexual relationship and a leader-follower relationship, based respectively on sexual and allelomimetic behavior. If all the different major systems of adaptive behavior were combined in pairs, some forty-five different kinds of relationships could theoretically be developed. In any one species only a few relationships are actually highly developed, but these vary so much from species to species that the study of the howling monkeys gives us only a preliminary idea of the possible kinds of social organization in the animal kingdom.

COMPARISON WITH OTHER PRIMATE SOCIETIES

The howling monkey society emphasizes three behavior relationships: care-dependency, leader-follower (allelomimetic), and sexual relationships, and these are usually well developed in other primate species as well. However, social organization may be radically different, even in other primate groups.

Baboons. For many years Carpenter's field studies of primate social organization stood alone in their field. Then, in the 1960s, there was a great revival of interest, partially stimulated by Sherwood Washburn and other physical anthropologists who were interested in the evolution of human family organization, and largely made possible by the increased funds made available in the United States for scientific research. Washburn and Irven DeVore led off with a study of savannah baboons, which are unusually interesting because they not only live in the general area in which man's ancestors developed, but also live in smaller-sized social groups and have the kinds of ecological relationships with their natural surroundings that most scientists agree must have been true of our very early human ancestors. These baboons live in savannah areas, grassy plains with occasional trees. A baboon troop of fifty to one hundred individuals wanders around in an area perhaps two miles in diameter, mostly living off tender young shoots of grass but also eating meat occasionally if a small mammal can be caught. The adult males are organized in a dominance-subordination relationship, with the most dominant male in the center of the group, and others spaced out around him. At the very edge are the youngest males, who are thus exposed to the maximum danger of predation, but who are also the most alert and active. If a potential predator such as a leopard or cheetah is sighted, they give alarm calls and all the males rush over and threaten the intruder, who usually retires. In this way the baboon troop can drive off any predators except lions, from whom the troop retreats to the trees. Nights are spent in certain sleeping trees, at which time the troop is most vulnerable to predators.

The dominance organization is a complex one. Both males and females may enlist the aid of others in their competitive interactions. Certain males consistently support each other, probably on the basis of a relationship established in early life. Thus a male's position in the dominance order depends both on his own strength and fighting ability and on his capacity to enlist the aid of others.

All females, which are much smaller than the males, are subordinate to all males, and develop their own weaker version of the dominance order. As the troop moves, they and their infants usually stay near the center, not because of any attraction to the dominant male, or any efforts on his part to keep them there, but because this is the safest place to be. There are no permanent male-female consortships, and male-female relationships based on sex are of a casual nature and confined to the brief periods when females are in estrus. The dominant male usually does a fairly large proportion of the mating, but females may also seek out others, and a female can pass from one male to another without competition between males.

The young baboons spend a large amount of time in play, chasing, rolling, and tumbling with each other. If one is hurt and cries out, the nearest adult male rushes over and threatens both parties. If a young male teases an old one, the latter gives chase, sometimes pursuing the young male up a tree and shaking the branches to which it clings until it becomes terrified, but never actually harming it. In this way young males learn to limit agonistic behavior to nonharmful activity, and learn their place in the dominance order without engaging in harmful violence.

Macaques. Rhesus monkeys have a social organization somewhat similar to that of baboons. They have been studied not only in India but in seminatural colonies established in Puerto Rico for scientific study. In the latter situation, where the individuals are identified and can be followed over a period of years, the finer social organization of a troop becomes apparent. There are no families such as the human nuclear family of father, mother, and children, but the offspring of a particular rhesus mother tend to stay close to her and to interact with each other in grooming and play. Furthermore, since the mother protects her young in any conflict, her offspring take on her rank in the dominance order.

As the troop moves, the young males and females are always in the lead. However, if the dominant male does not choose to follow, they return and take off in a direction indicated by him. While his relationship with respect to leadership largely consists of a veto power, in this particular animal society the most dominant individual also has functions of leadership.

Anthropoids. Man's closest relatives, the great apes, are more similar to him physically, but less so socially and ecologically. Gorillas, the most human appearing of all nonhuman primates, are in their ecological relationships specialized herbivores, like the hoofed mammals, and they have evolved in the direction of large size, eating huge quantities of vegetation, and developing large paunches. George Schaller, who studied them in their native rain forests, found that he could follow and observe them if he made the appropriate social signals. A direct stare is interpreted as a threat, and when two strange gorillas meet they normally direct their glances off to one side and eventually pass each other peacefully. Troops are small, consisting usually of one old male, two or three younger ones, and a few females with their offspring. There is almost no agonistic behavior within a troop. While the young animals generally give way to older ones, dominance organization does not appear to be important. On the other hand, the oldest male does act as a leader, determining the time of rising, the direction of travel, the time of resting, and that of nest-building. Nor is sexual behavior a prominent part of social activity. The sex organs are so inconspicuous that it is difficult to tell males from females, and sexual activity is infrequent. Consequently it is difficult to determine whether there is any group cohesion resulting from sexual behavior.

Chimpanzees, another forest-living great ape, show still another variety of social organization. Jane Goodall has studied one group of chimpanzees in the Gombe Stream Reserve in Tanganyika for several years, developing a method of quietly living in the same area as the chimpanzees until over a period of months and years they have accepted her as a harmless cohabitant. Unlike the baboons and rhesus, chimpanzees do not live in permanent troops. Rather, the animals living in the particular area form small temporary groups which wander over ranges of several square miles and may last for a few hours or days. These groups may be all males, all females, a combination between the two, or adult females with their infants. Only the mothers and their young infants stay together constantly. The most interesting observations concern not the social organization, which appears to be of a very loose nature except for that between mother and infant, but the way in which the chimpanzees live. They not only use tools, but occasionally kill small mammals and eat them, supplementing their herbivorous diet.

None of the above primate societies show true territoriality in the sense of a particular boundary that is defended. However, territorial behavior has been described for certain other primate species, such as the colobus monkey of South America and the vervet monkeys in Africa. In baboons, rhesus, and langur monkeys, the common organization with respect to space is that of a *core area,* inhabited by only one troop, with overlapping home ranges shared by two or more troops.

The kind of family organization seen in howling monkeys was originally thought to be a peculiarity of that species, since Carpenter had found something similar to the human nuclear family in gibbons, whose largest social group is an adult male, an adult female, and their immature offspring, a situation that actually results from a high degree of intolerance between adults of the same sex. However, all our new information about other primates indicates that the human type of nuclear family organization is the exception rather than the rule. Further, these field studies show, contrary to the hypothesis developed from observing nonhuman primates in captivity, that sexual behavior and sexual bonds are an unimportant part of group cohesion in most primate species. This raises the question whether sex behavior serves more than a simple reproductive function in any animal society.

Sex and social organization. Throughout the animal kingdom there is a great deal of variation in the amount of this behavior. It ranges from contacts which may not last more than a few hours or minutes in the entire life cycle of the animal, as in the nuptial flight of some of the social insects, through cases where there is frequent copulation throughout the early part of the breeding season, as in the song sparrow, to the situation of almost constant receptivity seen in adult human females. There is likewise a great deal of variability in the physiological causes affecting sexual behavior. In many animals sex behavior is primarily concerned with fertilization of the egg. Everything in the relationship is set up so as to insure fertilization and nothing else. For example, in many rodents the female exhibits sex behavior for only a few hours immediately prior to ovulation, which assures that the egg will become fertilized while it is in the best possible condition. Furthermore, sex behavior occurs at night, when the animals are most active and hence likely to meet males and at a time when there is presumably the least danger from natural predators. Once fertilization has taken place, there is no further opportunity for building up the sex relationship until the cycle of pregnancy is complete. An even more extreme example is seen in many of the hoofed animals. A female sheep shows sex behavior for approximately a day in the autumn of the year and if fertilization is completed will show no further sexual behavior until the following year. Under such conditions the sex relationship tends to form a relatively unimportant part of the animal's social organization.

The physiology of reproduction in birds requires more frequent sexual behavior. Most birds lay several eggs, and, unlike insects, which can store sperm for long periods, repeated fertilization is necessary. Consequently, mating behavior goes on throughout the laying season, and sexual behavior forms an important part of the male-female relationship. Combined with other social relationships, this creates a tendency in birds like wrens and blackbirds and ringdoves to form stable male-female relationships throughout the season and often over a period of several seasons, even though the males and females are separated during the winter.

There are also many species of mammals in which sex behavior is prolonged far beyond the needs of fertilization and in which the sexual relationship tends to become a stable and important part of a society. In wolves, which are the ancestors of our domestic dogs, the complete period of estrus in the female may last as long as a month or six weeks, with often-repeated sexual behavior. Under natural conditions, the males and females involved tend to stay together throughout the entire year. Similar relationships are set up in related carnivores. Fox-breeders often have great difficulty in getting a male to mate with more than one female.

In the anthropoid primates such as chimpanzees, there is a tendency for females to be receptive for long periods, so that they no longer have definite boundaries to the period of heat. The whole physiology of estrus has been altered, presumably in connection with this behavior. The phenomenon of menstruation is found only in primates. This evolutionary development can be understood if it is assumed that sexual behavior is no longer altogether concerned with fertilization but is an important part of social organization which tends to produce longer periods of attraction and, in some species, a more permanent relationship between adult males and females.

Leadership. In howling monkeys there is a tendency for the males to lead the females, although no one particular male is always the leader. In gorillas there is a definite male leader, and in many other primate groups the most dominant male has some leadership function. Exactly the opposite tendency is found in a flock of sheep organized on a natural basis, where the old female with the largest number of descendants consistently leads the flock. In true leader-follower relationships, the behavior concerned is allelomimetic, with both animals responding to each other but to an unequal degree. Among sheep the young lamb is constantly being called to the mother's side and rewarded by being allowed to nurse. Both mother and lamb are emotionally disturbed when the lamb is lost. It is possible that the emotions of anxiety and fear contribute to the tendency of the lamb to follow and that they make it a more lasting relationship than otherwise would be expected. In any case, the standard reaction

Fig. 27. Development of leadership in a flock of sheep. *Above*, first year. A female followed by her twin lambs, one male and one female. *Below*, second year. The same female leads, followed closely by her female lamb of the year before. Each of them in turn is closely followed by a new lamb. The male sheep born in the previous year now brings up the rear. The young lambs form a habit of following their own mothers, and this habit continues even into adult life, so that the oldest females tend to be the leaders of the flock. (Original)

of the lamb when threatened by danger is to run to the mother and follow her. As they grow older, the sheep slowly outgrow this, become more independent, and so eventually become leaders themselves. Since the older females have the largest number of offspring habituated to follow them, they become natural leaders when the flock moves. Wild herds of the red deer of Scotland show the same tendency for the older females to lead and the rest to follow.

There are very few authentic cases of consistent leadership in other species, in spite of folklore to the contrary. Leadership develops in flocks of ducks and geese, which like the sheep show an early following reaction. Some sort of leadership may exist in herds of wild horses, but no scientific study has ever been made of them. It is possible that more cases will be discovered in the future as investigators study animal societies more thoroughly, but we should remember not to confuse leadership with competition, in which two animals are racing toward the same goal, or with cases of dominance, where one animal drives or threatens another. As an example of the latter, the stags of the red deer each try to round up and hold together a small flock of females during the breeding season, butting the females when they try to get away. This is a far cry from the true leadership of the old females, which get out ahead of the herd and determine the direction of the group while the rest follow without any use of force.

What is the relationship between leadership and dominance? In a flock of goats there is some tendency toward leadership but not as extreme as in the sheep. The goat flock moves a good deal like the howling monkeys, with first one goat and then another taking the lead, but there is some tendency for the old females to get out in front. In contrast to sheep, the young kids are not kept constantly with the mothers but during the first two weeks of life adopt the "freezing" behavior similar to that of young fawns and are left behind while the mothers graze. This may account for the somewhat smaller degree of leadership. When a goat flock was tested for both dominance and leadership, it turned out that there was no correlation between the two and that an animal was both a leader and dominant only as often as might be expected to occur by chance. At least in this species, being dominant does not help in becoming a leader, and vice versa. The two relationships are apparently learned separately, and it is probable that they even conflict with each other, since one seems to be dependent on rewards and the other on punishment.

Fighting and social organization. The dominance-subordination relationship is highly developed in many birds, and one of the most elaborate examples is found in the sage grouse, a wild bird of the chicken family which lives on the western plains of the United States. These birds infect each other with parasites, and so they attracted the attention of a parasitologist, J. W. Scott. In studying the transmission of disease he came to discover a spectacular and beautiful example of mating behavior organized into a dominance order. No one had ever studied it before because it occurs only in the very early dawn in isolated mating grounds.

During most of the year the sage grouse is a quietly colored bird not unlike a Plymouth Rock hen, whose appearance blends well with its sagebrush environment. During the spring months, however, the male develops a flamboyant nuptial plumage which can be compared only to the court costumes of the Middle Ages. He grows long head plumes which can be raised or lowered, and his entire front is covered with brilliant white feathers which look like the ermine trim on a coronation robe. Behind, the tail feathers can be opened out into a gleaming, spangled fan. Before dawn the males in a given area assemble at a mating ground which is used over and over for generation after generation, where they go through elaborate behavioral posturing. The male spreads his tail, elevates his head, and with several movements takes in air so that two bright red patches of inflated skin appear in front of his white robe. Then he lets out the air with a dull "plop."

Each male establishes dominance relations with the other males on the ground, threatening them and buffeting them with his wings. The most dominant bird takes up his stand in a small area, and the subordinate birds form a ring around him at a little distance. If a strange bird comes near, they drive him off. These little circles of "guard cocks" around a "master cock" are found all over the mating ground. When the females arrive, they are admitted into the circle, and most of the mating is done by the master cock. This goes on over and over again during the moonlit nights of the early spring.

Mating behavior of the sage grouse is thus regulated by an elaborate dominance order. Dominance is not so obvious in perching birds like song sparrows and blackbirds, but, as each male bird sets up its territory by threats and fighting, it automatically becomes dominant in that area.

Any group of animals that can fight can set up a dominance order. Unlike most types of social organization, which tend to weaken and disappear in captive animals, dominance relationships in captivity can be strengthened or even created where none existed before. A herd of goats whose natural food is scattered over a field never fights over food as it grazes. If fed grain in a small area where the animals get in each other's way, they soon set up a dominance order in which the strongest animals have ready access to the food. Carnivorous animals like wolves, whose food supply is much scarcer, set up a definite dominance relationship as they devour their prey.

Thus dominance orders have a wide variety of functions, but

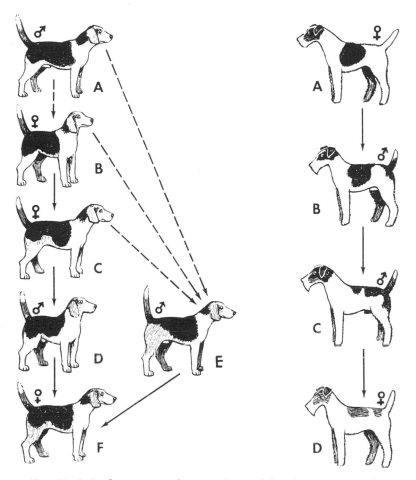

Fig. 28. *Left,* dominance order in a litter of beagles at 1 year of age. Complete dominance is shown in solid lines; incomplete in broken lines. In these non-aggressive animals only 4 of the 15 relationships showed complete dominance, and there was no dominance between *D* and *E*. The general picture is one of straight-line dominance, with rather weak relationships. *Right,* dominance order in a litter of fox terriers at 1 year of age. These aggressive animals show a perfect straight-line dominance order, with all dominance relationships definite and complete. Note that a female is at the top of the dominance order, although males are more frequently dominant over females in this breed. (Original)

they often result in the division of something which is limited in supply, such as mates, territory, or food. The most general effect of a dominance order is to regulate the space between individuals, as it does in a baboon colony.

Fighting can also be organized in other ways. The interesting habit of male howling monkeys which get together and roar at an outsider or a predator has its counterpart in many other primates. Baboons combine for mutual defense against predators, and rhesus monkeys may occasionally engage in intergroup conflicts, although attacks are never well coordinated. On the other hand, in herds of sheep and goats, the combats are always individual, each fight involving only two animals at a time. Musk oxen have a tendency to form a circle when attacked by predators, but they never combine against an individual of their own species. The combination of allelomimetic behavior with aggressive fighting that occurs in human warfare is very rare in animal societies. War may have some evolutionary basis in the tendency of certain primates to combine for defense, but war itself is a human invention.

Space and social organization. Agonistic behavior may have many different functions, but the primary one is the regulation of distance between individual animals. From his observations on wild domestic chickens described above, Glenn McBride has suggested that agonistic behavior produces the same effect as a physical "force field" which is counterbalanced by the effects of social attraction. The negative and positive forces tend to reach some kind of balance or equilibrium. When this balance is disturbed by animals approaching each other more closely than is normally permitted, the behavior of the individuals can be predicted on the basis of space, as the following example shows.

The brooding hens that McBride watched acted like solitary animals, staying apart from the roosters and other hens. Each hen always brooded her chicks in the same nest and usually fed in the same spot every day. Beyond distances of fifteen to twenty feet the hens paid no attention to each other. At distances of fifteen to twenty feet the hens raised their tails and began to scratch and cluck. Between fifteen and ten feet the hens raised their hackle feathers and the position of the tail opened to an inverted V. At eight feet the hens adopted the full posture of threat, with body and tail feathers fully expanded. When any fowl approached closer than this distance the mother hens were

15 — 20 ft.

10 ft.

5 ft.

Fig. 29. Space and social behavior in wild chickens. *Above,* Mother hen pays little attention to others at a distance, but begins to scratch and cluck as they approach within 20 feet. *Center,* at 10 feet she begins to expand her feathers. *Below,* at 5 feet she raises all her feathers, ready to charge. (Drawn from photographs by G. McBride)

likely to charge. Thus their behavior was highly correlated with the spatial relationships with other fowls, and its net effect was to maintain a distance of at least twenty feet from all other chickens.

The phenomenon of maintaining distance has been observed in many other social animals and the result is usually called *individual space,* the amount of space being fairly uniform for any particular species. Thus the agonistic behavior of the broody hens has the function of maintaining individual space, but not territory. However, in maintaining space they also exhibit dominance, just as barnyard fowls do.

Care-giving behavior and social organization. Some kind of care-dependency relationship is characteristic of any highly developed animal society. Among the social insects its role is so large that other social relationships are almost nonexistent. Worker ants spend most of their time gathering food for the young, feeding and cleaning them, and building elaborate nests for their protection. They extend this kind of relationship to each other. When two ants meet, they touch each other's bodies with their antennas, and one which has fed recently will regurgitate a drop of honey dew for the other. In this way the members of the colony take care of each other as well as the larvae. This relationship of *trophallaxis*, so characteristic of social insects, is a complex one, based on investigative, ingestive, and care-giving behavior.

A tendency for the care-dependency relationship to develop into *mutual care* is also seen in other animals. Monkeys and other primates often groom each other, going over one another's fur carefully and picking out parasites and bits of dirt. Other mammals, such as mice, comb one another's fur with their teeth and claws, and a single mouse in a cage often becomes very bedraggled compared to those kept in groups. Even horses will stand head to tail and whisk flies out of each other's faces.

Mutual care has obvious survival value for a species, but the care-dependency relationship involving the early protection of the offspring has even more basic biological and social importance. There is a tendency among many animals to provide care for their offspring over relatively long periods, and this raises the interesting problem of what happens to social relationships during the period of dependency.

SOCIALIZATION: THE FORMATION OF
PRIMARY SOCIAL RELATIONSHIPS

One of the simplest ways of studying early social relationships is to foster a young animal on another species. Many mammals and birds can be reared by hand, and it is a simple matter to exchange the eggs of birds which have similar nesting habits. Even in the social insects young larvae can be transferred from one nest to another.

Socialization in the ant. Worker ants ordinarily get along well with members of their own colony and will attack ants of a different species or of different colonies which attempt to enter the nest. As indicated earlier, this is not a hereditary trait, since

Fig. 30. Mutual care in chimpanzees. These adult animals are grooming each other, picking off particles of dirt. The basic type of behavior of both animals is epimeletic, or care-giving. (Drawing of chimpanzees at the Yerkes Laboratories of Primate Biology)

the slave-making ants are able to rear workers of other species and make them a part of the colony. An interesting series of experiments demonstrates what happens to a developing ant. If larvae of different species are taken out of their nests and raised together by the hand of a careful entomologist, no antagonism is shown between them. How then do the ants in natural colonies identify strangers? If an ant which normally arouses the antagonism of a different species is bathed in an alcohol solution and then in the body juices of the foreign species, it will not be attacked by the normally hostile species but will be attacked and killed by its own kind. Presumably a newly hatched ant quickly learns to associate peaceful behavior with the chemical taste and odor of its own colony. In this way the young ant apparently sets up a permanent social bond with members of its own anthill. The taste or smell of the colony is associated with mutual feeding or protection, and other tastes or smells with an attack. There is no opportunity for an ant to become socialized to another species, since death follows any contact with other colonies. Ant fights are not a matter of learning to avoid or dominate another individual but solely a matter of extermination. Occasionally one comes across examples of intercolony conflict on a large scale. The ants may go on fighting for a couple of days, leaving the ground littered with corpses.

Socialization in birds. Birds are much more easily raised by hand than are ants, and Lorenz and many ornithologists have described some remarkable instances of socialization to human beings. Lorenz emphasizes the importance of working with a wild species in which the instinctive mechanisms of behavior have not been allowed to become variable. Using the eggs of the greylag goose, he found that the newly hatched gosling would persistently follow any large moving object, including the experimenter himself. After a few days the goslings appeared to be strongly attached to human beings and did not respond at all to birds of the same species. Other related species form similar attachments, although in some cases the experimenter has to behave a good deal more like the normal mother bird. Quacking is required by some ducklings, and the experimenter may have to diminish his size by crawling around on the ground.

Lorenz emphasized the irreversible nature of this social process, for which he used the German term *Praegung*, usually translated as "imprinting." He thought that it was different from ordinary learning or habit formation because it seemed so sudden and irreversible. However, careful experimental work shows that imprinting is similar to the formation of a strong early habit. This habit, in conjunction with various behavioral mechanisms which prevent its modification, usually keeps a normally reared bird from becoming socialized to any other species. For example, several days after hatching, a mallard duckling develops a fear response which interferes with the following response and hence cuts out the possibility for further imprinting.

The chicken is of course not a wild species and shows much variability of behavior, but the chicks are easy to study and show a similar process of early socialization. Social behavior develops normally only when a very specific sequence of events is followed. For example, young chicks hatched separately show almost no reaction to each other until they have actually touched, after which they react strongly. Newly hatched chicks will often follow a person who clucks and walks away, but this tendency almost disappears by ten days of age. There is a limited time in life when this "following response" can be easily developed toward people, and this period may be shorter or longer in other species. In general, work with birds, and particularly with precocious birds like geese and chickens, indicates that there is a very short period in early life when primary socialization takes place.

Ducks and geese are a highly interesting group of birds. All are highly social, and many species are very successful, living in enormous numbers in streams, lakes, and marshes, as well as in ocean bays. They not only develop social organization within their own groups but are constantly exposed to contact with members of other and quite similar species. Consequently they face the evolutionary problem of avoiding mismating and hybridization. Ordinarily the various processes of socialization and attachment succeed each other in a regular fashion, culminating in each bird mating with its own kind, but even in wild situations hybrids are fairly common.

The total process is actually much more complicated than the original experiments with imprinting would indicate. For example, Friedrich Schutz has reared mallard ducklings in many different ways at the behavioral research station at Seewiesen in Germany. The station includes a small lake on which a large number of species of ducks and geese live under seminatural conditions and all mixed together so that each bird comes into contact with many species. Schutz first raised mallard ducklings only with their own kind. He then turned them loose on the lake, and the next year when they became mature, they mated only with other mallards. However, if he raised male mallards with another species, approximately two-thirds of them attempted to mate with the foster species, although only rarely with the same individuals with whom they had been reared. He also found that one-third of the mallards that were kept with their own species from one to three weeks after hatching, and then were reared with another species for the next five or six weeks, later attempted to mate with the second species rather than their own. This shows that the formation of attachments to particular individuals, as shown by the early following response, does not generalize to the species until some time later in development. Finally, the sexual, or mating, bond that is formed with a particular adult individual is only formed at the time of mating. Thus there are three different kinds of attachment processes.

To make things even more complicated, the processes of attachment vary from species to species. In those ducks like mallards, in which the males have a different plumage from the females, only the males develop a preference for their own species and thus show sexual imprinting. The females will respond if approached by males of another species that have been imprinted on mallards in their first few weeks of life, but are normally approached only by males of their own species. On the other hand, in species such as the Chilean teal, in which both the sexes are alike, both males and females become sexually imprinted. This difference between species has an obvious adaptive value since, in the case of the mallards, if female ducklings were imprinted on their mothers, they would later attempt to form sexual bonds with other females instead of males. In the Chilean teal, imprinting on the mother would still permit heterosexual mating, as she looks like a male.

Socialization in sheep. Animal breeders have long known that ewes will reject lambs which are not their own and occasionally will reject their own offspring when they have been handled and taken from the mother shortly after birth. Such rejected lambs are usually fed on a bottle and become very strongly attached to human beings. To accomplish this under experimental conditions, we took a female lamb from its mother at birth and raised it on a bottle for the first ten days of life. After that we put it out in the field where other sheep were grazing. We continued the bottle feeding and made no attempt to force the lamb into contact with the flock. When it approached the other sheep curiously, the mothers drove it away. The orphan followed an entirely different rhythm in its grazing pattern and had almost no contact with the flock, although they all stayed in the same small field. When it came into estrus, it submitted to being mounted by the rams but also stood still when caught by the human observer, presumably giving a sexual reaction. Even after a period of several years this sheep displayed a great deal of independence from the rest of the flock, not running when they were frightened and staying away from them on most occasions. In this case socialization is obviously associated with nursing, and the social behavior of the mothers prevents any possible resocialization at a later period.

Robert Cairns reared young lambs with dogs and found that they formed the same sort of attachments that a bottle lamb will form with a human caretaker. Later, such lambs would approach dogs instead of running from them, completely baffling a shepherd dog that attempted to herd them.

Another student of animal behavior, Collias, has carefully studied the behavior of sheep and goats at the time of birth. He finds that the lamb or kid will be accepted by the mother only within a very short period of approximately four hours. As in many birds, the process of socialization to the natural species is limited to a very short period, but in this case by the behavior of the mother rather than by that of the offspring. In Lorenz's terms the adult sheep is "imprinted" rather than the young animal. Practical goat breeders take advantage of this behavior by taking a kid away from its mother at birth. As the owner milks her, she reacts as if he were a kid. This attachment later makes the mother much easier to milk.

It is probable that the process of socialization in a lamb or kid is much more flexible than that in the mother. We raised a male lamb in somewhat the same way as we had the female. It had been rejected by the mother, and the owner left it with other sheep for four days while he fed it on the bottle. We then took it for a week and reared it with people before introducing it to our own flock. It was rejected by the flock in the same manner as our female lamb and formed much the same pattern of independent behavior. However, as it became sexually mature, it began to follow the females occasionally. As a result of this behavior it eventually became more closely associated with the flock than had the female. In normal sheep the process of primary socialization is limited to a very short period by the behavior of the mother, but the lamb itself is capable of developing this attachment and forming others until much later in life.

Socialization in the dog. Present evidence indicates that the dog was first domesticated at least ten thousand years ago, and careful anatomical comparison with wolves indicates that the wild ancestor was a small variety of wolf living near the fertile crescent of Mesopotamia, although the oldest known authentic dog skeletons actually were found in Idaho. Once dogs were domesticated, their use spread rapidly to all parts of the world and even to the continent of Australia, where they escaped from domestication. Probably because there were no competing placental mammals, they successfully established themselves and by isolation eventually became a separate species, the dingo dog. Elsewhere in the world the domesticated dog varied a great deal in the hands of different peoples, and its variability was undoubtedly accentuated by selection. The study of the socialization process in the dog is complicated by two factors: extreme genetic variability and the fact that socialization takes place readily with either dogs or human beings.

We have studied the developmental history of puppies in our laboratory in great detail and find that we can divide it into regular periods based on important changes in social relationships. These periods are roughly timed as follows, varying by a few days from individual to individual: a *neo-natal period* from birth until the eyes open at about 13 days; a *transition period* from the time the eyes open until the animal first responds to sound at about twenty days of age; a *period of socialization* which lasts from approximately three until twelve weeks; and a *juvenile period* from this time until the animal is first capable of mating behavior, which may occur at any time from six months to more than a year of age.

During the neo-natal period the behavior of the puppies is limited to reflexes and simple behavior patterns concerned mainly with nursing, elimination, and warmth. These infantile responses are adapted to a situation in which the puppy is completely dependent on maternal care. The transition period is one of rapid development of sensory, motor, and psychological capacities, as well as patterns of social behavior. At its end the puppy is capable of many complex forms of behavior and consequently begins to be independent.

The psychological development of the puppy is strongly protected from the environment during the first two periods, both by the nonfunctioning sense organs and limited motor capacities, and by the consequent limitation on the ability to make discriminations and form associations. While some degree of attachment takes place as early as ten days, the puppy only becomes fully capable of forming rapid attachments at about three weeks.

The electrical activity of the puppy brain, as measured by the electroencephalograph, runs parallel to the development of outward behavior. A newborn puppy shows almost no brain waves at all and little differentiation between waking and sleeping states. A large difference between sleep and wakefulness appears almost exactly at the beginning of the period of socialization, and the brain waves take their final adult form between seven and eight weeks, close to the time of final weaning.

Most of the important patterns of social behavior appear in some form early in the period of socialization. Besides nursing, which has been going on from the very first, the puppies exhibit playful fighting and by six or seven weeks of age may show allelomimetic behavior as they run in a group. Occasional sex behavior appears in the form of mounting. The only types of behavior not found are mature sexual and care-giving behavior, which do not appear until the adult stage. Since social behavior determines social organization, all the basic social relationships begin to be formed in this early period.

From the accounts of trappers and those who have raised wolf puppies, a very similar kind of development takes place in the wolf. It is apparently quite easy to socialize a wolf puppy to people if it is taken shortly before the eyes are open, which would of course be before the start of socialization. A few weeks later the wolf cubs have turned into wild animals, snarling and snapping viciously when handled, and few people would have the temerity to take them on as pets.

We have done several experiments on the socialization of puppies with respect to human beings. The degree of socialization can be measured by testing the puppy for the presence or absence of fear reactions toward people. If the puppies are raised under kennel conditions with relatively little human contact until five weeks of age, the majority of them show fear reactions which mostly disappear within the next two weeks if they are handled often. If the puppies are taken from a litter at three or four weeks of age and raised by hand, they show no fear reactions at five weeks. However, if the puppies are allowed to run wild with no human contact until twelve weeks or so, they become increasingly timid and are almost impossible to catch. A puppy of this age can still be socialized if caught and forced into close human contact by confinement and hand feeding, but it will always tend to be somewhat timid of human beings and less responsive than the animals socialized at the early period.

There are also important genetic differences between individuals and between breeds. Some strains in breeds like the African basenji show a relatively large amount of initial wildness which is greatly reduced by handling, whereas the cocker spaniels usually show a small amount which is never reduced. It is probable that in the wolf the period in which socialization to humans can take place is very short and that it has been considerably extended by selection in most of the dog breeds.

Another measure of socialization is distress vocalization following separation from the individual to whom the attachment is made. Beginning at about ten days of age, near the onset of the transition period, a young puppy will start to vocalize if it is either left alone in its home nest or if it is separated from its mother and litter mates and placed in a strange room. These reactions reach a maximum at three weeks of age and continue at the same high level until eight weeks, when the rate begins to go down. A beagle puppy will average 120 vocalizations a minute in a strange room and about half of this rate when isolated in a familiar room. As soon as the puppy is restored to his home room and familiar companions, the reaction stops almost immediately.

These facts indicate a mechanism by which an emotional attachment is formed. As soon as the puppy develops the sensory capacity to distinguish between familiar and unfamiliar animals and places, it shows an almost reflex emotional response to separation. This should act as an internal reinforcing mechanism, punishing the puppy during separation and relieving the unpleasant emotions on reunion. From our knowledge of the process of learning, this mechanism should, with repeated separations and reunions, result in the development of a strong degree of social motivation or attachment.

Attachment is thus dependent upon familiarization, which may be defined as recognition that certain stimuli have been received before. Cairns and Werboff introduced young puppies during the period of socialization to strange companions, such as rabbits. They reacted to separation from the rabbits by vocalization after only two hours of contact, and showed the maximum degree of emotional disturbance after only twenty-four hours of association. The process of familiarization thus takes place very rapidly at this age.

Understanding the nature of the attachment process makes many things clear. There is no reason to expect that the process of familiarization and the emotional reaction to separation should cease in older animals. In fact, as pet owners know, the emotional reaction to separation in an older dog may be even more severe and long-lasting than in a puppy, as we would expect from the strengthening of motivation through reinforcement. What does happen in later life is that a developing fear response to the strange prevents any long contacts with strange individuals or places and hence prevents the opportunity for familiarization. Similarly, another form of agonistic behavior, that of attacking and threatening strange dogs, particularly with reference to the dog's own territory, prevents new contacts. Finally, the very fact of becoming strongly attached to certain dogs or persons automatically prevents long contacts being made with others.

The attachment process itself is almost impossible to upset in a very young puppy. It does not require hand feeding, nor even active interaction. Puppies that have been fed entirely by machine still form attachments to persons who stay with them, and a puppy will learn to run faster toward a passive human being than toward one that actively holds and pets it. Furthermore, dogs that are actually punished for their contacts with human beings still form attachments to them.

The best explanation of these facts is an evolutionary one. For a highly social and dependent animal like a puppy, survival depends upon forming and maintaining social contact. If a young wolf cub becomes separated from the adults, it will die in short order. Therefore, if a wild wolf cub were inhibited from forming an attachment to other wolves when it was accidentally hurt by them, it would simply not survive. The whole process then, is a basic one for survival.

There are indications that the same general process and mechanisms account for the socialization process in other animals. Even in ducklings and chickens, separation is followed by distress vocalization, food rewards are unnecessary, and painful stimulation does not halt the process of forming an attachment. Likewise, a period in which attachments are rapidly and easily formed is followed by a time when fear responses inhibit new contacts.

Socialization in wolves. Naturalists have long known that a wolf cub taken from the litter before the eyes open or soon after and raised by hand will make a delightful pet, if somewhat more difficult to manage than a domestic dog because of its larger size and great strength. Recently Benson Ginsburg and his co-workers have found that even wild-caught adult wolves can be socialized. The process was discovered during experiments with drugs that were supposed to produce a taming effect on the wolves. One of the control experiments was for a laboratory assistant to approach and attempt to handle a wolf that has not been drugged. Much to his surprise, Ginsburg found that some of these wolves began to react in a friendly and tolerant fashion.

On this basis he developed a system for socialization involving the wolf being isolated in a large cage. Every day the experimenter would go into the cage and sit down quietly and read a book for an hour or so. At first the wolf reacted with extreme fear, trying to climb the walls, salivating, urinating, and defecating. After several weeks the wolf eventually calmed down and one day would make a positive approach toward the experimenter. At this point it was essential to do the right thing, neither frightening the wolf nor acting fearful. If the experimenter acted in a confident and friendly manner, the wolf accepted him and from that time also would react in a friendly fashion toward other human beings. The relationship was different from that developed with a wolf cub caught at an early age, as the adult wolf socialized in this way is neither subordinate nor highly dependent. It does not react toward the person as it might toward another wolf, as is characteristic of animals socialized at an early age, but rather as if the human were an interesting and rewarding object for social interaction.

Further experiments with wolf cubs showed that the attachment formed in early life is not necessarily final. Young cubs socialized to people and then returned to a wolf group in the zoo reacted as adults like their companions who had never had this experience. Apparently there is a later period when the earlier attachment is solidified and becomes permanent.

Socialization in primates. The process of socialization in monkeys and apes has not been studied in such detail, but, as we would expect, there is a period early in life when primary social relationships are formed. Arboreal primates like the howling monkeys show a long period in which the infant is constantly carried in the mother's arms or rides on her back. Until it is at least two years old a young monkey cannot keep up with the adults as they travel through the trees. Ordinarily the mother has only one offspring at a time, giving an opportunity for the development of a very strong relationship between mother and young.

Experiments in which baby chimpanzees are removed from their mothers show that the babies can readily be socialized to human beings if obtained within the first two years of life and that this becomes increasingly difficult later. There may be some mechanism of timidity which causes this increasing resistance to socialization, which in adults is complicated by aggressiveness. One of the striking things about socialization in a primate is the long period during which it may take place, as contrasted to a few hours or days in the birds and a few weeks in the dog, although this period may still be relatively short compared to the total life span of the animal.

The most thorough set of experiments on the socialization of primates was done by Harry Harlow and his associates at the University of Wisconsin primate laboratories. Compared with chimpanzees, rhesus monkeys are born in a much more mature state. Under natural conditions they are carried by the mother and cling to her fur as soon as they are born. In attempting to raise monkeys away from their mothers, Harlow found that they could not survive in barren cages with a flat floor unless they had some object to which they could cling. He therefore designed a model mother which was a mother-sized object covered with soft cloth, and the baby monkeys, clinging to this substitute mother, survived beautifully. In order to test the effect of food rewards, Harlow devised an uncomfortable model made of bare steel wire to which was attached a nursing bottle which was the only source of food. Later, given preference tests, the monkeys preferred the comfortable model in spite of the fact that it had never provided food. As with the experiments in dogs, and indeed as with the earlier experiments with imprinting in ducklings and chicks, food rewards are unnecessary to the attachment process. As pointed out in an earlier chapter, female monkeys reared on model mothers made poor mothers themselves, rejecting their infants and punishing them when they cried. However, the infants still crawled toward their mothers, indicating that the attachment cannot be inhibited by punishment. Thus the attachment process as it has been studied in rhesus monkeys, dogs, and even in birds, appears to be fundamentally similar and depends primarily not on external reinforcement but on the opportunity for long continued contact plus repeated internal reinforcement resulting from separation and reunion.

Present evidence from studies with children indicates that the process of attachment is basically similar to that in the nonhuman animals described above. We would predict, on this basis, that human beings would form attachments to whatever objects and persons with whom they have prolonged contact. Robert Zajonc has found objective evidence that this rule holds true even with such things as words and letters. When human subjects are asked to rate their preference for words and letters, the results follow closely the relative frequency with which these occur in the language. For an English speaking person, "e" is a much more attractive letter than "z," for example.

Critical periods for primary socialization. All highly social animals seem to exhibit a process of socialization, and this can be divided into two parts. *Primary socialization,* which usually takes place relatively early in life although not necessarily at birth, determines the group of animals to which an individual will become attached. This group is usually of the same species, but socialization can be experimentally transferred to other species. *Secondary socialization* to other animals and groups may take place in later life, as in the formation of sex relationships. However, primary socialization often limits very strictly the kind of secondary socialization which may take place. Most species seem to have behavioral mechanisms which make it difficult to form attachments to dissimilar individuals once primary socialization has taken place. These mechanisms differ from species to species. In some, primary socialization must take place within a few hours, and with others, it may take place over a period of years. There is also evidence that much hereditary variability affecting primary socialization can occur within a single species.

All this leads to the conclusion that critical periods exist in the process of socialization. Primary socialization tends to be self-

Communication in a red-winged blackbird.
The male bird sits on a post in the middle of
its territory and sings as a warning to intruders. This "song-spread display" is a
communicatory form of agonistic behavior.
(Photo by R. W. Nero)

Vocalization in mammals is generally inferior to that in birds, as is demonstrated by attempts to
train them to talk. Viki, a chimpanzee reared with the attention usually given a child, never mas-
tered more than three vocalized words. Here she is pronouncing the word "cup" for her foster
mother, Mrs. Hayes. She always put her hand over her face while saying this, which added sign
language to the vocal word. The word itself was a loud whisper, with the "c" pronounced like the
German "ch." (Photo by Keith Hayes)

Period of socialization in the puppy. During the early part of this period, which extends from approximately 3 to 12 weeks, a puppy will form strong new attachments in less than 24 hours. The optimal time to take a puppy from its mother and litter mates and adopt it as a house pet is between 6 and 8 weeks of age. It is ready for final weaning from the breast and still capable of rapidly forming attachments to people.

Vocal communication in the dog. This sonogram of a puppy's response to being separated from its mother and littermates pictures a series of rapid high pitched barks emitted at the rate of 5 per second. It is easy to locate the direction from which such short, sharp and similar sounds are produced, with the result that a lost puppy is easily located.

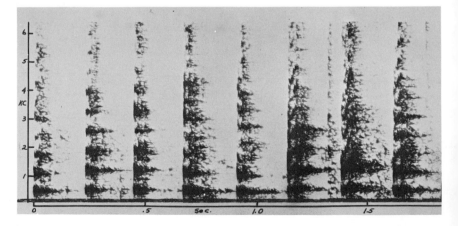

limiting and also to limit the formation of secondary social relationships. The time at which any new social relationship is normally formed tends to be a critical period for the socialization of the individual.

These general findings are related to the human problems of antagonism and mutual tolerance. Some kinds of birds raised by hand will give mating reactions to the human species and reject their own kind. On the other hand, puppies which have contact with both dogs and people during the period of socialization become attached to both species. This raises the question: Could improved methods of socialization in human infants increase the variety of people whom they can later tolerate and to whom they can successfully adjust in adult relationships?

When we look at an animal society as a whole, we find a striking correlation between the final organization typical of the species and the social development of the young individual. In the dog and wolf the mother stays constantly with the young in the first few weeks, then begins to leave them for long periods just at the beginning of the period of socialization. This means that the strongest primary relationship is developed with the other pups rather than with the mother. This in turn lays the foundation for the formation of the pack, which is the typical social group of adults. Similarly, the behavior of female goats in leaving their newborn offspring while they feed is correlated with the relatively weak leader-follower relationship in these animals. We may conclude that one of the important factors that determines the structure of an animal society is its type of social development and the process of socialization that goes with it.

HUMAN BIOLOGY AND THE BASIS OF SOCIAL ORGANIZATION

Biologists have long been aware that a high degree of social organization exists in certain species of animals, and the first ones to come to their attention were the social insects and man. Only within comparatively recent years has it been realized that almost all animals show some degree of social behavior and social organization. Even the most primitive and solitary animals tend to form groups for mutual protection under certain conditions, and their behavior is to that extent social, while the higher vertebrates show complex social relationships which often surpass those of the insects.

As we have seen, social behavior can be differentiated into social relationships by either biological or psychological means, and more often by a combination of the two. The psychological process itself depends upon two kinds of biological abilities. One of these is the ability to express the particular kind of social behavior which goes into the relationship, and the other is the ability to organize social behavior through learning. Animals differ a great deal in these basic capacities. At one extreme are the termites, which show a high degree of biologically determined social organization. In some species there are as many as six different castes, each with a special behavioral function in the colony and each different in physical appearance from the rest. Their social behavior is differentiated by heredity, and there is little evidence that they have any power to organize it through learning. As might be expected in a society which is so strongly affected by biological factors, the termite colony as a group has many of the properties of an individual organism and can be called with some accuracy a "super-organism." The behavior of other social insects seems to be much more affected by experience, with somewhat greater flexibility in their social organization. For example, the worker bees perform different functions around the hive at different ages. However, there is still a tendency for an individual's place in the social system to be strictly correlated with its physical structure.

A popular pastime among entomologists is to compare insect and human societies, usually to the disparagement of the latter. In a well-running insect society, everything is smoothly organized, with no crime, neglect of children, juvenile delinquency, or unemployment. This raises the question: Could human beings ever achieve a society as stable as those of insects? The answer rests on another question: What is the biological basis of human social organization?

In the first place, human beings start by being far less biologically specialized than insects. They conform to the typical vertebrate pattern of three biological types—male, female, and young, with no further subdivision into castes. Among vertebrates there is a widespread tendency for the males to specialize in fighting and dominance relationships and for females to concentrate on care-giving behavior and the development of care-dependency relationships with the young. There is also a general tendency toward differences in sex behavior. However, there are many exceptions to these generalities, and the actual behavioral roles of the two sexes vary a great deal. In some birds, such as pigeons and doves, there is no external anatomical difference between the sexes, and the differences in sexual behavior are relatively slight. In fish it is typically the male which builds the nest and guards the developing eggs, while the female has the sole function of depositing the eggs at the appropriate time. Herd animals go to the opposite extreme. Almost the entire care of the young is done by the females. The males, with their superior size, strength, horns, and aggressiveness, show a great deal of specialization for fighting.

We can find the same wide variability among man's primate relatives. As we have seen, there is only a moderate degree of contrast in the sex roles of howling monkeys. Neither sex fights very much, and both take some care of the young. Male and female gibbons are even more alike. The females are very nearly the same size as the males and, unlike the howlers, both males and females are extremely belligerent to members of their own sex. The result is that the gibbon social group never includes more than an adult male and female with their developing offspring. Baboons have strongly contrasting sex roles, the males being much bigger and more belligerent than the females.

It is not possible to reason that because man is a primate he should show a particular basic type of social organization. There is too much variety in primate social behavior, and man is not closely similar to any other living primate species. If there is anything which is genuinely characteristic of all primates, it is that both sexes show at least some of all the various kinds of social behavior exhibited by the species, and this corresponds to what we know of human behavior. It is probable that there is some biological differentiation of behavior between the human sexes but that it is not as extreme as that in some primates. There is a great deal of variability in behavior between members of the same sex and considerable overlap between the two.

Man is capable of a high degree of differentiation of behavior on a psychological basis, and human beings go far beyond the strictly biological organization found in some insect societies. With our great development of psychological organization, is it possible for us to achieve the extreme division of labor and uniform control of behavior that we see in insects? The answer, as nearly as we can give it at the present time, is that extreme specialization and rigid control of behavior are not well suited to man's biological nature. There is a tendency toward variability of behavior implicit in the process of learning which cannot be completely eliminated even by the most rigid habit formation. A human being confined to too narrow a social role will not show his full range of capability. There is every evidence from the clinical psychologists that such a person will be thwarted, frustrated, and unhappy. Man's biological nature equips him for developing a wide variety of complex social relationships. Each of these can be developed with a great deal of individual variability and is subject to all sorts of modification by learning and experience. The ideal human being is one who develops a variety of social relationships suited to his individual needs and capacities, and the ideal human society is one which is based on these relationships.

CHAPTER 9

Communication
The Language of Animals

Every socially organized group of animals must maintain contact between its members. In some groups, such as army ants, this is done by actual physical contact, but by far the greater number of animal societies, like those of fish and birds, keep track of each other while the individuals move around freely and at considerable distances. The method used is partly dependent upon the sense organs of the species concerned and partly on the way the group members stimulate each other through some sort of motor activity. The process may be thought of in terms of stimulus and response, but it also involves the idea of communication.

We human beings are so used to the idea of verbal communication that we are apt to assume that all animals can or should communicate in this way. Kipling with his *Jungle Books* is only one of many imaginative writers who have envisioned the remarkable things that a man could do if only he knew the language of other animals. In the usual plot, the animals have a secret language very much like human speech which the hero is somehow able to learn, and the rest of the story deals with the remarkable results. In recent years scientific study has shown that many animals do have secret systems of communication, some of which resemble human language in many ways. However, what the animals say to each other is usually not what a romantic writer might imagine.

COMMUNICATION IN BIRDS

Birds have great advantages for behavioral study, and it is not surprising that they have yielded fruitful results far in advance of other groups. They do not compete with man as much as most

213

of the mammals. A great number of bird species have been able to survive and live in natural societies, even in areas where civilized men—and of course this includes the scientists—live in great numbers. Besides being plentiful, birds are easy to observe. In contrast to many of the other vertebrates, which are able to survive close to man only because they live in holes or creep around by night, the vast majority of birds fly around in the daytime in a conspicuous manner.

Many years ago birds were chiefly studied by naturalists and occasionally by very amateurish naturalists. These "bird watchers" sometimes developed a kind of cult whose sole object was the quick identification of birds on the wing, and as such they often became figures of fun. Nevertheless, the facts that they gathered were usually correct, and some of the most interesting advances in the understanding of bird behavior have come from gifted amateur ornithologists.

One of the earliest activities of the bird watchers was to describe bird sounds in an accurate and objective way. In the older bird books there are many attempts to render bird calls in terms of nonsense syllables and the musical scale. The king rail is reported to give a loud startling cry of "Bup bup bup!" whereas the mourning dove has a sad—to human ears—call which may be written "Coo-o-o, ah-coo-o-o, coo-o-o, coo-o-o." It usually turns out that any particular species has at least six or seven different calls. Most of these can be easily interpreted as indicating danger, distress, hunger, presence of food, and so on. Among the perching birds there is one special call whose significance is not so obvious —the song of the males. In some birds, like song sparrows and thrushes, it is very musical, and in others, like catbirds, it is sung with many variations. The ornithologists duly noted that the song was usually a definite way to identify a species even when the bird itself was not visible, and that it was heard most often in the mating and nesting season. It remained for Eliot Howard to point out that the typical bird song is actually a territorial cry which serves as a warning to other males.

It is expected that more discoveries will be made in the future, mainly because of modern technical advances. Nonsense syllables and musical notation give at the best only a very rough approximation of the actual sound. With modern sensitive microphones and tape recorders, it is possible to record bird calls accurately

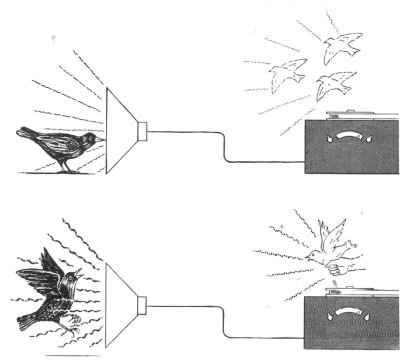

Fig. 31. Birds may be either attracted or repelled by recordings of the appropriate bird cries. Starlings leave the area in which a record of their distress call is played. (Based on work of Frings)

and to play them over and over again. A more elaborate technique is to transform the sound through an oscillograph to a graphic representation, or "picture" of the bird call. This is usually transformed into a sound spectogram (sonagram) in which the frequency of sounds is graphed against time. Or it may be recorded more simply as an oscillogram, a photograph of sound-wave motion which chiefly represents duration and loudness of sounds. In this way, characteristics can be measured which are not apparent to the human ear. Once this is done, it is possible to analyze bird sounds and their effect on other birds in great detail.

There have been some attempts to use the new techniques to talk to birds in their own language. One of the most spectacular of these concerns the starlings, European birds which were imported and turned loose in the United States, where they promptly became a great pest. They are particularly troublesome in the autumn and winter, when they come into many towns and cities to roost in trees and on the surfaces of public buildings, making raucous noises and spreading their droppings liberally over buildings and passers-by.

Hubert Frings observed that when a starling is caught it utters a loud shriek of distress. He carefully recorded its cries, procured a sound-amplifying truck, and went into a small Pennsylvania town much infested with starlings. He toured their roosts one evening with the sound amplifier turned up as high as possible and emitting starling shrieks of alarm in hitherto unheard-of volume. The effect on the human inhabitants was not recorded, but the starlings left the village for good. It would seem that there are considerable practical applications in using the language of birds. Subsequent experience shows that the method is not always completely effective, as the starlings sometimes return. We would expect that they would eventually habituate to the sound, or even learn to discriminate between the recorded distress call and natural sounds. Certainly birds can discriminate between sounds coming from different birds, as the following example shows.

J. B. Falls and his colleagues at the University of Toronto have done many such experiments in the field. The ovenbird is familiar to those who live near the northern woods as the bird that makes a loud noise that sounds like "teacher—teacher—teacher." This is the territorial song of the male and consists of about ten repetitions of the teacher cry in about three seconds. Individual birds vary their delivery enough so that even human ears can identify them; and as Falls's experiment shows, the birds themselves discriminate between the recorded calls of their neighbors and those of strange birds. In the spring after the territories had been well established, Falls set up his speaker at various points in the birds' territories and played various recordings to them. The resident bird in each paid almost no attention to the calls of his neighbors but flew over and responded vigorously to the songs of strange birds, singing back and moving around the speaker as if searching for the source of the sound. If the strange song was played back repeatedly, the resident bird rapidly became habituated to it and paid no further attention, but if the speaker was moved to a different location, he again became excited and responded to it.

Thus, the territorial behavior of the bird, which at first seems mechanical and stereotyped, actually involves individual recognition of songs and discrimination between them. The information conveyed to the resident bird, however, is simple; namely, that a strange bird has entered his territorial boundaries at a particular spot.

All modern evidence agrees with that of the early ornithologists: that any bird species has a limited number of sounds and that each of these has a very limited meaning to other birds. However, a great many problems remain unanswered. Why do bobwhite quails, which apparently have no regular territory, give their characteristic call during the morning and evening? Do they keep in touch with each other in this way, and what effect does it have on other birds? Why do mocking birds and catbirds give not only the characteristic songs of their own species but also those of other birds, and what effect does this have upon a blackbird that is trying to defend his territory?

Some birds are good imitators of human speech, and still another way of approaching the problem of animal language is to try to teach them to talk our own. The ill-fated attempt to communicate with the horse Clever Hans is a disappointing example, but birds are much more promising subjects. Many experiments in rearing birds apart from their own kind indicate that at least some species learn the typical territorial song from their parents. English sparrows raised with canaries will learn the canaries' song. Baltimore orioles raised in isolation develop a song quite different from the wild birds, and when other orioles are raised with them, the young ones sing in the new fashion. It follows that some birds have considerable gifts for vocal imitation.

Certain species, like parrots and parakeets, are well known for their imitative powers, and crows and mynah birds are also good performers. However, teaching a bird to talk is not an easy process. The best way is to obtain a bird very young and socialize it to the human species. With a great deal of attention and by being rewarded when it repeats human speech, it will eventually learn to say several recognizable words and even sentences. Sometimes the bird says them in appropriate circumstances, but no one has been able to teach a bird to recombine words in new phrases. A bird's use of human language appears to be little more than a complicated trick which can be used to earn a reward.

Birds also communicate with each other in other ways than sound. The remarkable display behavior of the sage grouse is not an isolated instance. A great many species have characteristic patterns of behavior which act as visual signs or signals to their fellows. In most cases the nature of these sign stimuli seems to be largely fixed by heredity, and the essential part of the behavior may be some very small but characteristic part of the pattern. On the other hand, the response given is not so firmly determined and may be transferred by learning and socialization to another stimulus, as when a hand-reared bird responds sexually to the human hand.

In general, our present evidence indicates that communication in birds, either vocal or through visual signals, conveys rather simple information, the extent of which is largely determined by heredity. No one, however, really knows what the highly vocal birds are up to under natural conditions.

COMMUNICATION IN MAMMALS

Echolocation in bats. Bats are nocturnal animals which usually live in caves during the daytime and fly out in the evening to catch their food. Unlike nocturnal birds such as owls, which have very large eyes and are well adapted to seeing in dim light, the bats have extremely small eyes. In spite of this, they fly in and out of their caves in dense throngs without collisions, catch insects on the wing with precision, and can fly at top speed around a crowded room, avoiding obstacles with a last-minute flip of the wings. It looks as if bats had some sort of special apparatus which human beings do not possess. Donald Griffin and his associate, Robert Galambos, were able to demonstrate the nature of this ability by a series of ingenious experiments.

Bats have very large ears, and when these are plugged up so that they cannot hear, the animals lose their power to avoid objects when flying. The next step is to put a sensitive microphone in a room with the flying bats. Modern electronic equipment, including the oscillograph, shows that the flying bats continually emit bursts of high-frequency sound which are inaudible to human observers. Although some of the noises which bats make can be just barely heard, most of them are far beyond the range of the human ear. If the bat's mouth is plugged instead of its ears, it again is unable to avoid objects; while if obstacles such as very fine wires are strung up before a flying bat, it is unable to detect them even if in full possession of its faculties. Griffin concluded that the bat orients itself by giving out short bursts of high-frequency sound which echo back from any solid object. It is thus able to avoid obstacles and locate food. Very fine wire will not give sufficient echo for a bat to detect, and hence it flies into such

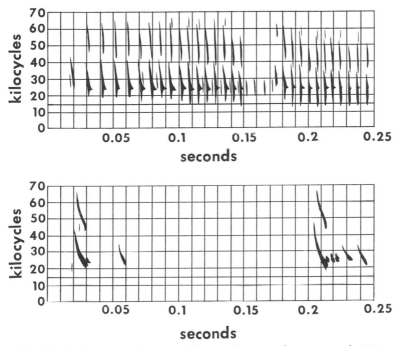

Fig. 32. Analyzing vocalization of bats with a sound spectrograph. Using electronic equipment, it is possible to "see" sounds which are far above the range of the human ear. In the graphs the vertical scale shows the number of vibrations in kilocycles. The upper limit of human audition is about 20 kilocycles, or 20,000 vibrations per second, so that only a small part of bat sounds can be heard. The horizontal scale is time in fractions of a second, so that the whole graph covers the noises made by a bat in a quarter-second. *Above*, noises made by a bat pursuing an insect. The marks concentrated around 25 kilocycles are the principal noises made by the bat, and the very thin lines following them are the echoes, which are easily visible after the first burst of sound. The marks at the higher range of sound are the harmonics. *Below*, sounds made by a bat flying high above the ground. The bat made two principal sounds, one about 0.03 seconds and the other at 0.21 seconds, each followed by echoes. (Sound spectrograph analysis by D. R. Griffin)

objects as if they were not there. Fortunately for bats, such things are rare in nature.

This discovery raises several interesting questions. Do bats communicate with each other by means of supersonic sounds? As the bats fly out of the caves, they must hear each other as well as the echoes of their bodies. And what of the many small nocturnal rodents which make high-frequency noises, some of which are audible to the human ear? The possibility exists that field mice and deer mice call to each other soundlessly in the night. We know that young rodents of many species make supersonic noises that stimulate their mothers to care for them.

Porpoises and sonar. During World War II experimenters invented a device for detecting enemy vessels which they called sonar. This consisted of making a standard noise, or "ping," and recording the echoes which came back. As the naval scientists listened to these water sounds, they discovered that the ocean is a very noisy place, and other scientists began to investigate the sounds that fishes and other marine animals produce. W. N. Kellogg, who had long been interested in problems of animal communication, began to work with the bottle-nosed dolphins, or porpoises, and discovered that these animals had been using sonar long before human beings.

Porpoises themselves are fascinating animals. Like most other members of the whale family, they show no fear of man and frequently exhibit what appears to be a friendly interest in human swimmers. Like wolves, these animals are preadapted for domestication, and they adjust very well to life in aquariums, as anyone knows who has visited the spots where they are placed on exhibition. They have large brains and are easily trained to perform elaborate tricks.

The sounds that they produce are a series of high-pitched clicks, far beyond the human audible range, plus a variety of other sounds. It is the former that are used in echolocation and they are very similar to those of bats. Placing opaque cups over the eyes of porpoises, Kellogg showed that they could still locate target objects. A porpoise will work very hard to obtain fish, and porpoises will not only learn to avoid objects but also distinguish between different-sized fish when given no visual cues.

Since the porpoise makes so much noise and is so well equipped to analyze sounds, this raises the question of whether these animals use sound for communication. They can undoubtedly detect the noises made by other porpoises, but finding out how much information these convey is a difficult process, partly because water is such a good conductor of sound. It travels about five times as fast in seawater as in air, and very little of the energy is dissipated. Underwater explosions have been detected by human devices over a distance as long as 10,000 miles, and theoretically a sound made in one spot could be heard around the world if there were no intervening land barriers. If porpoises are communicating, it is possible for them to do this over very long distances and proportionally difficult to find out experimentally what other porpoises are receiving the communication and how it affects their behavior. J. C. Lilly is convinced that porpoises have a brain large enough to be capable of language learning, and has attempted experiments with teaching porpoises to imitate human language. The results are hard to evaluate, since the sound-producing apparatus of the porpoises is so different from that of a human being. However, it is certain that porpoises emit a variety of sounds similar to emotional communication in other mammals. Other species belonging to the whale order also make a variety of sounds, and the "Song of the Humpbacked Whale" achieved considerable popularity as a recording. The sea lion, a marine mammal belonging to the order Carnivora, also makes a variety of sounds, but is primarily visual in its orientation and does not use echolocation.

The expression of emotion. Very little progress has been made since Darwin's day on the analysis of mammalian social communication. Many mammals, like the dog and wolf, have a variety of postures and behavioral patterns which stimulate other members of the same species, but most of these seem to convey only simple information about the emotional state of the animal. For example, when a dog holds its tail very erect and wags it slowly, it is a sign of a dominant attitude. When two strange males approach each other in this fashion, a fight frequently follows. A crouching posture with the tail held low and wagging rapidly is associated with subordination.

Some rodents, like prairie dogs and other ground squirrels, vocalize a great deal, but when analyzed these sounds seem to be quite similar to the territorial calls and warning cries of birds. Though the sounds are frequently repeated, there is little variability in the patterns of vocalization. Unlike the situation in birds —it must be emphasized that it has not been so thoroughly studied in mammals—these sounds and signals do not seem to produce any specific reactions, and the mammals have to learn what they mean. For example, a male mouse which is approaching a stranger and hesitating between attacking and running away often vibrates the tail very rapidly, producing a rattling sound on the wall of the cage. This produces no observable effect on the behavior of an inexperienced mouse, although presumably he could learn eventually to recognize it as a danger signal.

Our results with dogs gave somewhat similar results. Barking in dogs and wolves has the function of being both an alarm signal and a threat, although many house dogs use it as an attention-getting device as well. We supposed that if this sound conveyed any precise information, young puppies should give a uniform reaction to it. Actually, puppies will show a startle reaction to any loud noise, and we prepared a series of tape recordings that included barks from several dogs, including the mother, and as a control sound having somewhat similar acoustical properties we recorded the noise made by hitting a board on a table. We also included a completely different sound, a pure tone. As it turned out, the puppies paid no more attention to the barks of the mother than those of strangers, and about the same to the sound of the board. Their greatest response was to the completely novel sound, the pure tone. They definitely did not give any uniform alarm reaction to any of the barks. Only about one-sixth of their responses were fearful, the rest being alerting and searching.

In other experiments we recorded the distress vocalization of puppies and played them back to the mother. If the puppies were outside the room, she became very anxious and attempted to get to them. However, if they were in the room with her, she merely took one look at them and paid no further attention. Apparently dogs react to these signals very much as a human being might, reacting first with attention and then responding in terms of the whole situation rather than to the signal alone.

Incidentally, the general acoustical properties of a bark are very similar to those of the sounds produced by porpoises and bats in echolocation, except that the basic sound in the bark has a low frequency. In barking, a dog produces a series of short, sharp sounds very similar in pitch and volume, and rapidly repeated. While these are not used for echolocation, the properties of the bark make it very easy for any other dog or human being to pinpoint the exact direction from which they come, as changes in sounds coming a fraction of a second apart are easy to discriminate. In contrast, the songs of many birds vary greatly in pitch and loudness, giving a ventriloqual effect which makes the bird itself very difficult to locate, as anyone who has attempted to get a glimpse of a singing bird in a leafy tree can testify. As Peter Marler found, the alarm calls that many birds give to the presence of a hawk are even more difficult to localize, usually consisting of a high, thin whistle that fades in and out and gives the listener few cues of directionality. A similar ventriloqual effect is produced by the vocalizations of coyotes. Unlike the sharply different howls and barks of wolves and dogs, these are a mixture of barks, howls, whines, and yaps, with the result that the vocalizations made by a pair of coyotes sound like a pack of forty calling from many different directions.

Teaching mammals to talk. Efforts to teach mammals human language have been notoriously unsuccessful compared with birds. A trained dog can learn to discriminate between as many as seventy-five or one hundred individual words, but efforts to teach dogs to talk have never produced more than one or two faintly recognizable sounds.

The chimpanzee, with its brain and motor apparatus so much more similar to those of human beings, would seem to be a better prospect, but chimpanzees are still relatively poor at making sounds. One study of these apes in captivity lists thirty-two different noises which could be identified by the human listener, but this is not much better than the dog with its various kinds of barks, howls, and whines. Chimpanzee noises are certainly no more complex than bird calls and range from "Ho-oh!" to express alarm, to a barking "Uh, uh, uh, uh" while eating, which evidently indicates satisfaction. Chimpanzee vocalization is explosive and consists mostly of vowels. All of these sounds seem to express some kind of emotional reaction.

The various attempts to raise chimpanzees completely away from their kind and socialize them like children have never resulted in the chimpanzees learning to talk human language. The Kelloggs raised a chimpanzee with their own child, and Viki was raised by Dr. Keith Hayes and his wife, Cathy Hayes, as an only child. Even with this concentrated parental attention, she was able to learn only three human words by the age of three years. She made very few babbling noises compared to human babies and stopped doing this by the age of five months. Her foster parents then tried to teach her to "speak" for a food reward, and by another five months she learned to grunt on command. After this they manipulated her lips to produce the word "mama," which she soon learned to do for herself. Then they found that they could imitate some of Viki's sounds by using a hoarse whisper and tried to use this noise as a base for further words. Viki herself was very poor at copying any kind of human noises, in spite of the fact that she readily imitated other actions. The lessons went slowly, and by the time she was two-and-a-half, she could say "papa" and "cup." Eventually she learned to use them in the proper way and would whisper "cup" when she wanted a drink of water.

It is possible that through improved technique another chimpanzee could be taught a bigger vocabulary in less time. However, Viki's foster parents felt that she had very nearly reached her limit in acquiring a total of seven words by the age of six. By this time she could communicate a great deal more to them by gestures and picture cards. At best, Viki's performance in learning a language is far behind those of human children, who often have vocabularies up to hundreds of meaningful words by the age of two and are already combining them into sentences.

Like Viki, chimpanzees in their native forests communicate at least as much by gestural signals as by vocalization. Allen and Beatrice Gardner have therefore raised a female chimpanzee in much the same way as Viki, but instead of trying to teach her to talk have attempted to teach her the standard American sign language used by the deaf. Washoe (named after Washoe County in Nevada where she was reared) was perhaps not an ideal subject, since she was not adopted until the age of eight to fourteen months. Twenty-two months later, at about the age of three, her foster parents had taught her to use thirty-two different signs, some of them in combinations, such as "Give me (one sign) drink, please." While the number of signs learned in this time was no more than equal to the total number of vocalizations that chimpanzees naturally use, the results are far better than any results with attempting to teach vocal language. Washoe has now moved to the University of Oklahoma where she is continuing her education with considerable success, and where other chimpanzees are being taught by the same methods.

All the evidence indicates that the chimpanzees and other mammals that have been unsuccessfully trained to talk have a poor sort of vocal ability. The larynx and other motor apparatus is there, but they seem to have very little voluntary control over it. It is difficult for them to repress noises when emotionally excited, and even more difficult to produce the noises voluntarily in the absence of emotion. Compared to birds, mammals are relatively silent and make a small variety of noises. It is possible that experiments with some of the noisier primates, like the howling monkeys, might give better results, but for the present we must conclude that man is a unique mammal in at least one way —his vocal powers.

THE LANGUAGE OF BEES

The honeybees have roughly the same relationship to other members of the order Hymenoptera as man does to other primates. Their social organization is vastly more complex and permanent than that of other insects, and they have excited the interest and imagination of scientists and writers from the earliest historical times. Most of the other bees and wasps start new colonies each year from a single fertilized queen that lives over the winter, while in the case of ants each colony is started by one mated pair and has to be built up anew in every generation. The honeybee society is never reduced to a single queen or pair. New colonies are started by an old queen with a swarm of workers, leaving some of the workers with a young queen behind in the old hive. In both the old and new colonies there are always mature individuals present, and so there is a much greater opportunity for learned behavior to be passed along.

Furthermore, there appears to be a division of labor which is not based on purely biological factors. There are three kinds of adults in the hive: the queen, or fertilized egg-laying female, which can be produced at various times by special feeding; the male drones, which develop from unfertilized eggs; and the workers, which are genetically like the queen but have not been as well fed. It is this latter class which shows a great deal of differentiation of labor within itself and which suggests the possibility of psychological differentiation such as is found in some of the higher vertebrate societies. The ordinary worker bee during the height of the summer season may live five or six weeks. When bees are marked for individual study, it is found that young ones work mostly around the hive, feeding the larvae, ventilating the hive, and acting as guards around the entrance. They usually do not go on field work until ten days of age, but they may go earlier when only younger bees are present. Once a worker bee gets started on a particular occupation, it will stay with it for several days, and, when field work is started, a particular bee which carries water, pollen, or nectar to the hive will keep doing this almost indefinitely without any change of occupation. At the same time a bee can show considerable flexibility of behavior. If bees are fed at a particular time of day, they soon come to the feeding spot only at the correct time. When the feeding is stopped, the bees return for a few days at the right time and then stop coming. The implications are that bees are capable of a rather high degree of adaptation, habit formation, and memory, and that these abilities make possible the psychological differentiation of social behavior.

It has long been suspected that bees are capable of informing each other about the location of food supplies. If honey or other desirable food is left out where bees can find it, the usual result is that one bee comes along, locates the food, and then flies away. Shortly afterward, a large number of other bees appear. How does the first bee take back the information to the hive? Bees make very few noises and these chiefly with the wings. The queen, under certain circumstances, makes a sort of quacking noise; and when the hive is disturbed the normal buzzing sound of the bees' wings rises in volume to almost a roar. These sounds are quite limited in variety, but they do play a part in communication.

One of the difficulties of studying bees is that individuals cannot be conveniently taken into a laboratory for detailed study. A beehive is a complete and functioning social system, and behavior can be understood only while the animals are a part of this system. In addition, bees are not domesticated in the sense that many vertebrates are tamed and partially socialized to humans. Bees are not even dependent on human care. Bee colonies can by various subterfuges be lured into living in patent hives where they can be more easily robbed of their honey or studied scientifically, but they can, and frequently do, escape to the wild and take up residence in hollow trees or even in the hollow frames of buildings. The honeybee is not a native of North America, and when the first European colonists brought beehives with them, the bees escaped and rapidly colonized the whole continent, so that the later human pioneers found bee trees already established in the primeval forests.

A major discovery concerning the system of communication among bees came as a result of a lifetime of patient study by the German animal behaviorist Von Frisch. When a foraging bee returns to the hive from a rich source of food, it not only deposits its load of pollen and nectar but crawls around on the honeycomb in a way that at first looks like random movement but is actually a definite pattern of behavior which is not the same for every bee. As it moves around, the bee wags its abdomen from side to side, and Von Frisch gave this phenomenon the happy German title of "Schwanzeltanz" or "waggle dance."

Von Frisch did most of his first experiments by placing a feeding station close to the hive. When the bees came back to the hive, they performed a simple circular movement which he called a "round dance" and in which there is no wagging of the abdomen. In later experiments he moved the site of the feeding station farther away and found that the dance changed to the "wagging dance" between 50 and 100 meters. The bee regularly goes through a figure-eight motion, stopping in the middle to wag the abdomen, and this is repeated several times. He then counted the number of turns made and found that there was a definite inverse relationship between the number of turns and the distance to the feeding station. At 200 meters the bees averaged approximately eight turns in fifteen seconds, while at 500 meters they averaged only six. After 1,500 meters, or approximately one mile, the curve rapidly levels off, and beyond 5,000 meters the curve is practically level. This is close to the greatest distance to which a bee will forage.

Von Frisch also found that a bee always headed in a definite direction when wagging her abdomen. In some cases a bee would head upward while doing the wagging part of the dance, and when this happened, the source of food was always in the same direction as the sun. When the bee headed downward, the food was always away from the sun.

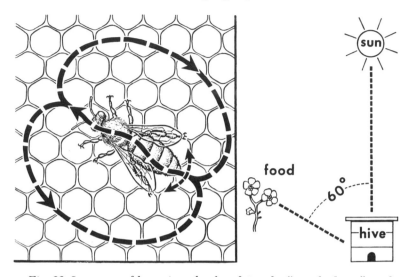

Fig. 33. Language of bees. A worker bee doing the "waggle dance" on the vertical surface of the honeycomb in an experimental hive. The bee moves upward at an angle of 60° to the left from the vertical while wagging its abdomen (the food is about 60° to the left of the sun). At the top of its run, the bee stops and makes a circle back to its starting point, turning in the opposite direction each time so as to make a figure **8**. The bee makes 8 circles in 15 seconds (the food is about 200 meters away from the hive). With this information, the other bees are able to locate the food rapidly. (Based on Karl von Frisch, *Bees: Their Vision, Chemical Senses, and Language,* 2d ed., © 1971 by Cornell University Press)

The bees do not always head straight up or straight down, and the angle is related to the relative position of food and sun from the hive. For example, a bee dancing with its head 60 degrees to the left of the vertical means that the food source is 60 degrees left of the sun. Von Frisch found that bees were able to locate food very accurately within 15 degrees.

Apparently the bee, like some ants, uses the sun to orient itself. This phenomenon has been called the "sun compass." This brings up the problem of what bees do on cloudy days. The hive is usually not so active, but the bees do go out foraging and are able to tell the direction without difficulty. The clue to this behavior came after a long and complicated series of experiments during which it became apparent that bees could orient themselves as long as some patch of blue sky was visible. Von Frisch found that the bees in this case were able to orient themselves because the degree of polarization of light in the sky is related to the position of the sun. This means that on a partly cloudy day with the sun obscured, the bees can still work and communicate with each other. Even on a completely cloudy day, the bees can locate the sun if sufficient ultraviolet light penetrates the clouds. Because their eyes are sensitive to ultraviolet light, bees can see the sun when it is invisible to a human observer.

This is still not the whole story. Adrian Wenner has done additional experiments that show that bees transmit information by means of odors and sound as well as by the dances. A successful foraging bee comes back to the hive carrying the odor of the food and flowers with which it came into contact. As it dances, it gives off regular pulses of sound during the straight run of the figure eight, the number of pulses being correlated with the distance from the hive. These sounds and dances arouse the other bees. Inexperienced bees detect the odor, fly downwind from the hive, and then search for the proper combination of odors. Experienced foragers have a memory of where odors have previously occurred, and quickly check these locations. Wenner's experimental results lead him to believe that while the human observer can read the language of the dances, bees themselves ordinarily do not. It is probable that bees receive information concerning the location of food in more than one way, or perhaps in a combination of ways.

In another bee species, sound is an important part of communication. In colonies of the stingless bee of South America (*Melipona quadrifasciata*) returning foragers begin to hum when they return to the hive, producing these noises at equally spaced intervals but making longer and longer sounds depending on how far away the feeding place was. Esch and Kerr trained stingless bees to visit a feeding station ten meters from the hive and marked all of the successful foragers. Then they took the food away until bees no longer visited the site. At this point they played a tape recording of sounds made by bees returning from a distance of zero to thirty meters. Both experienced and inexperienced bees immediately turned up at the feeding station. Ordinary honey bees also emit sounds whose length is related to the distance of the food source while they are doing the waggle dance, but on a somewhat different scale. Since the stingless bees do not show dancing behavior, it is probable that communication first evolved with respect to sound and that the dancing is a more elaborate system of communication developed by the honey bees, which still use the information from sound and odor signals as well. The final answer with respect to the powers of communication in the honey bee is still to be obtained.

All these complicated activities of bees lead us to conclude that they are able to do something which had long been thought was possible only for human beings, namely, that one bee can tell another the location of something else, an abstract fact about a third object which could not be seen or otherwise perceived by either. Bees really do have a secret language. It is not a verbal language but a system of signals, and the knowledge of it does not permit us to communicate with the bees, because only a bee can give the right kind of signal.

PHEROMONES

In insects. The use of food odors in communication among bees is only one way in which animals communicate with each other by means of scent. In the Queen butterfly, a close relative of the large orange and black Monarch butterfly, Lincoln Brower studied courtship behavior by rearing females in the laboratory, thus making sure that they were unmated, and then releasing them one at a time in the vicinity of wild males in southern Florida. As a female flies off, she is pursued by a male who attempts to get near her and extrudes two organs called "hair pencils" from the rear of his abdomen. The hairy tips of these organs can be spread out into two flower-like circles of hairs. With them he brushes the head of the female and so releases a sticky, sweet-smelling substance whose odor is strong enough to be detected by the human observer at close range. In response she settles down on a nearby twig or leaf, the male settles upon her and, after stroking her alternately with his left and right antennae, copulation takes place, after which the pair separate and fly off.

In these butterflies the chemical substance brushed on by the male acts as a primary stimulus for the female's becoming quiet and assuming a position which makes mating possible. The active ingredient has been identified as a ketone carried by a sticky dust and unless so stimulated females will not mate. Brower concludes that it is a true aphrodisiac, although only for butterflies.

In other insects odors can be used to transmit information. In the fire ants studied by E. O. Wilson, the workers forage individually, but if a worker finds a piece of food too heavy to carry, it returns to the nest, laying down a trail with an odorous substance from a gland in its abdomen. Other workers are attracted by the odor and return to its source, with the result that many ants are soon pulling and hauling on the piece of food. The odor fades away quickly, with the result that ants are attracted only for a short time and for short distances.

Many species of ants, when alarmed, secrete substances which excite other ants to run in a circular and zig-zag fashion. While in this excited state, especially if near the nest, they will attack anything that they come into contact with, sometimes even attacking the ant that has released the odor in the first place.

Substances such as these have received the name *pheromones*, in an analogy with the term hormones, which act as internal chemical messengers. The pheromones are ectohormones, chemical messengers acting outside the body. They not only function as primary stimuli, or releasers, as in some of the above examples, but may also produce physiological effects in the same manner as internal hormones. These changes, in turn, may directly affect behavior.

Pheromones in mammals. One of the most striking effects of odorous stimulation occurs not in insects but in vertebrates. Working with problems of fertility in laboratory house mice, H. M. Bruce discovered that females that had recently mated and ordinarily would have become pregnant did not do so if they were exposed to strange males. Later experiments showed that the effect of pregnancy-block was more likely to occur when the strange male was most unlike; that is, a male from a different strain is more likely to produce the effect. Females do not need to come into contact with the males, or even see them. An odor, or pheromone, produced by the strange male is the essential stimulus. Presumably this works by the female's smelling the odor. This stimulus is transmitted through the brain to the hypothalamus, which responds by secreting hormones that prevent implantation of the eggs. Thus the female goes into another reproductive cycle without producing young. There are some fascinating implications regarding the effect of this mechanism on population growth and the distribution of genes. An incursion of strange males would have the effect of reducing fertility, but if any male remained in contact through the next estrus cycle he would no longer be a stranger and would be more likely to be a successful father than the original male, thus furthering outbreeding.

This latter possibility is increased by the fact that still another pheromone secreted by male mice has the effect of inducing and speeding up the estrus cycle in the female. A third pheromone is produced by females. Wesley Whitten found that when female mice are confined together in the absence of males, the estrus cycle is suppressed. Again, this is caused by an odor passed from female to female, since cutting the olfactory nerves prevents the effect.

Thus there are two main effects of pheromones. Those producing physiological effects are called *primers,* and those inducing direct behavioral effects are classified as *primary stimuli* or *releasers.* Among mammals, the latter result is not nearly as precise and mechanical as in insects, and the more general term "signaling pheromone" is preferable. Among these, the substance produced by the chin glands of rabbits is a good example. The European rabbit, the wild ancestor of our domestic form, was introduced into Australia, where it rapidly multiplied, with correspondingly great destruction of the vegetation. R. Mykytowycz studied a colony of marked individuals under seminatural conditions. As wild European rabbits do, they dug warrens and soon established dominance orders within each sex, with all males being dominant over all females. The area around each warren became a territory, and in this area the males of the warren marked objects by rubbing them against glands under their chins. While moving in their own territories, rabbits repeatedly sniff and chin objects, but once a territorial border is crossed they immediately assume a posture indicating fear and wariness. While the territorial boundaries are not as precise as those in birds, the effect of the substance is to signal to any strange rabbit that the area is occupied and that he might be attacked if discovered.

The "mysterious" powers of animals. Enough has been said to show that remarkable discoveries have been made in the field of animal communication, and many scientists are hopeful that this will lead to still further discoveries. However, a word of scientific skepticism should be inserted at this point. There are real mysteries in animal behavior, things which we do not understand and which presumably will be eventually explained, given patience and hard work. This is no reason for offering mystical explanations or for endowing animals with supernatural powers. On the other hand, we should not be too skeptical. Animal behavior is not supernatural, but animals occasionally turn out to have powers which are definitely superhuman, such as the bees' ability to detect polarized light and the ability of many mammals to detect high-frequency sounds. In exploring new phenomena, we need to draw a line between explanations which are supernatural and those which are merely superhuman. The original experiments on the "sun compass" of ants showed that they were able to orient themselves even when inclosed by a box so that the sky was visible but the sun was not, and the experimenter could only theorize that the ants were in some way perceiving the stars. This explanation verges on the supernatural, since it is hard to see a physical way in which insect eyes could filter out all the interfering light from the sky and still leave that of the stars. The true explanation appears to be that ants, like bees, are able to detect differences in polarization of light in different parts of the sky, which is an ability not found in human beings.

ANIMAL LANGUAGE AND HUMAN LANGUAGE

Students of human behavior are fond of pointing out that in at least one respect humanity is unique in the animal kingdom. Human beings have a highly developed system of verbal communication, and it is often argued that there is nothing else like it, and that its possession sets man distinctly apart from all other animals. However, the facts show that animal systems of communication possess in rudimentary form most of the basic characteristics of human language.

In the first place, human communication is basically vocal, but there are numerous birds and mammals which also have some sort of vocal communication. Secondly, a great deal of information regarding the emotional state of the human speaker is conveyed not so much by his words as by his tones of voice. As far as we now know, the vocal communication of birds and mammals consists almost entirely of conveying feelings and emotions. Human language has been said to be unique in that it can be used to inform people about other objects and events outside the speaker;

but this can be done also in the communication of bees. Human language can be used as an instrument to produce effects on other individuals, but so can the cries of other animals. A person confronted with a locked door calls to someone to open it, while his pet dog barks to produce the same effect.

Human speech can also be used as a system of symbols for solving problems, and there is no good evidence that other animals make use of verbal symbols in this way. Chimpanzees and other mammals can solve problems whose complexity would seem to indicate the use of symbols, but the actual explanation may be some sensory perceptual process analogous to what we call visualization. Another outstanding characteristic of human language is that it can be converted into nonvocal symbols, such as written words, and thus can influence activities at far distant times and places. And yet many of the carnivorous animals, like wolves, mark certain posts and trees, and others that come along days later may react to them.

The one characteristic of human language which appears to be unique is the fact that it is largely independent of heredity. Any human being can learn any kind of language, and there is a great deal of variety and flexibility possible. It is very likely that the communication system of bees is governed almost entirely by heredity, but if it could be shown that any part of their system is passed along from individual to individual by learning rather than by biological heredity, it would have almost all of the essential characteristics of human communication. In short, we may conclude that human language has evolved from capabilities which are present in a large variety of lower animals and that the chief differences lie not so much in possessing a different kind of basic ability as in the degree to which language has been developed and its importance in social organization.

The phenomenon of communication in social animals increases the amount of organization developed by a group and the distances over which the organization can extend. The behavior of any one bee belonging to a particular hive is related to that of the others, even though the workers may be scattered over an area with a three-mile radius. Social organization and communication thus lead to the formation of organized populations, whose structure is also affected by another basic result of animal behavior, the tendency to become attached to particular localities.

CHAPTER 10

Behavior
and the Environment

Any visitor to the seashore has seen limpets clinging tightly to the rocks in tide pools. If you can pry one loose, you find that underneath the shell these small mollusks resemble their relatives the snails, having a broad foot with a head and tentacles at one end. In the ordinary course of its life, a limpet crawls slowly around on a rock covered by high tide, grazing on the small marine plants which grow there. At low tide it clamps itself tightly to the rock, using the broad foot as a kind of suction cup. The success of this maneuver depends upon a tight fit between the shell and the rock. When the English animal behaviorist E. S. Russell marked limpets for individual recognition, he found that they always returned to the same place at low tide and that, as the shell grew, it took on the shape of the particular piece of rock which was the home of the limpet.

This sort of behavior is very common among animals. Far from moving around at random over the face of the earth or even moving to adjust to changing environmental conditions, the great majority seem to have definite preferred localities around which they spend most of their lives. These spots are usually the places in which they were born and grew up. This behavior suggests that in addition to the process of socialization there may be an even more widespread phenomenon of becoming attached to a particular spot on the earth's surface. This behavioral process may be called *localization*, used in the sense that an animal becomes related to a particular place, rather than in the psychological sense of orientation with regard to sound. With localization we begin to deal with behavior on the ecological level of organization.

THE ECOLOGICAL ORGANIZATION OF BEHAVIOR

The problem of the effects of behavior goes beyond social organization and communication. As well as being related to particular groups of animals of its own kind, the behavior of an individual affects that of other species and becomes related to particular areas of the earth's surface. When we consider facts such as these, we are dealing with phenomena on a new level of organization which is usually called ecological. The most important relationship between animals of one species and another is that of food-getting. All animals obtain their food by eating plants or some other species of animals, and each in turn is usually the food for still other species. From this viewpoint, an animal species is a link in a food chain. A minnow eats small invertebrates and in turn is eaten by larger fish. We have already covered most of the behavioral aspects of this type of organization in our study of ingestive behavior, and the detailed economics of interspecies relationships belongs to the study of ecology. On the other hand, the organization of behavior in relation to particular places has a great deal of behavioral interest and results in the formation of a new unit of organization—the population.

Primary localization: Attachment to a home area. Attachment to a particular geographical spot is almost invariably first formed early in life. More than sixty percent of a group of house wrens banded as nestlings were later trapped within a mile of their original nest. White-tailed deer, which are now becoming so common in the American countryside, can also be caught and tagged as young animals and their subsequent movements followed. Even such large animals as these stay most of the time within a half-mile radius of the place in which they grew up. The deer will not move even when food becomes scarce; and if they are forcibly driven to other places they soon come drifting back.

Even more spectacular examples are seen in some of the lower animals. Salmon, which live in the sea during adult life and return to fresh water to spawn, almost invariably come back to the same river in which they grew up. In one experiment 11,000 tagged salmon were recovered in the parent stream, and none elsewhere. They are able to detect differences in the chemical composition of the water of different rivers and seldom if ever make mistakes.

Learning to recognize the special type of river water must take place in early life before the seaward migration.

Nest-building and shelter-seeking. Some adult types of behavior also tend to keep animals in particular localities. Animals which build nests and use them to care for the young center their behavior around this locality. Many birds come back to the same area each year for nesting purposes. In one observation 31 percent of male wrens came back to the same nest box, and another 53 percent settled within a thousand feet. Some birds, like eagles, build their nests in exactly the same spot year after year. In social insects like ants and wasps all the activities center around the spot in which the nest is built. Simple shelter-seeking behavior is probably another factor in keeping animals in particular localities. Bats are dependent upon caves for shelter during the daytime and for hibernation in the winter, and they show a tendency to return to the same cave.

Orientation and homing. The fact that an animal's behavior becomes localized to a particular spot implies that it must be able to find its way back if it goes away. How this is done is well known in some species but remains an almost complete mystery in others. The problem is particularly difficult in the case of birds, some of which fly over extremely long distances during migration. Various experiments have shown that birds are expert at finding their way back to their homes, and this ability has been used in a practical way for centuries in the case of the homing pigeon. A properly trained pigeon released in a strange place takes off and flies around the release point for about three and a half minutes, then appears to orient himself on a rather straight line for home, flying at an average speed of about eleven miles per hour. In general, performance is very good over short and very long distances and rather poor in intermediate distances. The scientific evidence shows first that genetics plays a part: homers are much better than common pigeons. Second, the birds improve with practice. Most important, birds perform very poorly if they are raised in confinement, especially if their cages have opaque walls and they cannot see the horizon. If they are regularly waked up at midnight and placed in darkness at noon, they go off course by 90 degrees, indicating that their direction finding, like that of many other birds, is based on an internal biological clock.

Birds have an excellent sense of sight and frequently fly at considerable heights. At comparable altitudes human beings can see a hundred miles in any direction in clear weather. It should be correspondingly easy for birds to recognize landmarks from long distances. The best way to check these ideas is to study bird behavior directly, and the large clearly visible sea birds can be followed in a slow-flying light airplane. Using this technique, Griffin found that gannets which had been taken four or five hundred miles away from their homes did not take direct paths back but did considerable circling around after being released, indicating that they might have been using visual landmarks.

It is also possible that birds make use of what we often call a "sense of direction." In people this is based partly on visual landmarks but also on a sort of kinetic memory of the turns which have been made. A person with a good sense of direction rarely gets confused, even at night and when very gradual turns are made, as in the case of a railway journey. This ability is probably dependent upon the semicircular canals of the internal ear, and it is possible that birds have it developed to a much greater degree than people.

Even on these bases it is difficult to explain the feats of such migratory birds as the golden plover which flies across the Pacific Ocean from Alaska to Hawaii with no landmarks en route. The suggestion has been made that such birds are charting their course by the sun or stars in the same way as human navigators. To do this with a ship, it is necessary to have an accurate clock and a sextant to obtain the angle of the sun, since its position changes with the time of day. It would not be too difficult for a bird to observe the angle of the sun with its eye while it kept its head level in flight, and, as for clocks to keep track of the passage of time, it is possible that birds have some sort of internal timing mechanism.

One of the difficulties of testing these mechanisms by which birds orient themselves is that of eliminating various objects which might give the birds cues, as well as that of following the birds in their long flights. For this reason, orientation in the penguins of Antarctica provides an ideal opportunity for experimentation, provided one can arrange transportation. The continent is covered with snow and ice which provide no reliable landmarks for the birds. Since penguins are flightless, they can move only relatively slowly over the Antarctic wastes. In soft snow they leave plainly marked trails behind them, or they can be followed and observed directly. John Emlen and Richard Penney took Adelie penguins from Cape Crozier and released them elsewhere, sometimes as much as 200 miles away. No matter where they were placed, the birds always chose a direction that would have led them directly to the sea had they been inland from Cape Crozier; that is, in a line perpendicular to the coast at that point. That the birds were guiding themselves by means of a sun compass was shown by the fact that on cloudy days their paths were less consistent. If the sun were completely obscured, the birds might wander in any direction. While the sun was shining, the birds moved in very straight lines. In order to do this, they must have been constantly correcting the information received from the sun with the aid of some sort of internal biological clock.

When birds from a different location were set free near Cape Crozier, they moved in a direction different from the paths of the local birds, but one which would have led them to the sea on their own side of the continent. When some of these new birds were left three or four weeks near Cape Crozier, they apparently reset their internal clocks and chose the same paths as the local birds. Thus, the theory of celestial navigation for orienting in birds receives strong support, incredible as it sounds to human observers who can only achieve such feats with elaborate scientific apparatus developed over periods of centuries.

Birds that migrate by night must navigate with reference to the stars rather than the sun, and one hypothesis is that such birds have a "genetic star map." Stephen Emlen hand-reared indigo buntings in order that they would have no contacts with adults, and found that if they were never allowed to see the sky at nighttime they were unable to select the normal southward direction of migration when exposed to the artificial skies of a planetarium. Those that had been exposed to the planetarium skies when young headed in a normal southward direction, while those which were exposed to a fictitious planetarium sky headed in what would have been a southerly direction according to the fictitious sky. These experiments show that the skill for using the stars in navigation has to be developed, but they do not explain why the buntings always take a southerly direction in their autumnal migrations, even without cues from other birds.

It is also probable that different species of birds use different methods or combinations of methods for orienting themselves, and that no one single explanation will fit them all. For example, recent experiments with homing pigeons indicate that these birds can guide themselves by sensing the direction of magnetic lines of force, if they are forced to fly at night or under conditions of continuous cloud cover, when the sun is invisible, although ordinarily pigeons use the sun as a compass.

Biological clocks. Many animals show definite daily and seasonal rhythms in their behavior, always performing certain acts at particular times of the day or year. In some cases, these are obvious direct responses to changes in external stimuli, but others are largely independent of outside influences. These internal "circadian rhythms" could serve as the biological clocks used by birds in their migration.

Seasonal rhythms are produced in different ways. The fall migrations of birds have long been known to be related to declining day length, and the northward migration in the spring, together with the increase in size of the reproductive organs and accompanying mating behavior, can be produced by artificially lengthening the day in a gradual fashion. Similarly, the breeding season of Basenji dogs, which normally takes place in the autumn of the year, near the time of the autumnal equinox, can be advanced to early summer by artificially shortening the day. Presumably, such effects are produced by the brain through visual stimulation. The brain in some way detects these rather small daily changes and conveys the information to the pituitary gland, which then either produces hormones or quits producing them.

A more remarkable case is the timing of the hibernation of the woodchuck. These animals become active at the appropriate time in February even when kept under constant laboratory conditions. This means either that an internal biological clock is able to function accurately over many months, or that the woodchucks are being stimulated by external stimuli of which human investigators are not aware. Further progress on this problem will probably come when someone is able to locate a spot in the brain where a lesion will stop the biological clock.

Migration. As an extension of the tendency of animals to stay located in one spot, some species have two homes which are regularly visited according to the season. The mule deer in parts of the

Homing behavior of limpets brings each animal back to the same spot at low tide. This assures a tight fit of the conical shell to the rock surface, not only because the shell conforms to the same rock surface as it grows, but because the rock becomes grooved by the friction with the shell edge. In this picture the largest limpet occupies its rock scar neatly, but several smaller limpets occupy spots that show the scars of larger predecessors. Many animals have a tendency to become localized, or attached to particular spots, as well as becoming socialized to particular animals. (Photographed at Trevone, England, by R. Buchsbaum)

Type of open bivouac formed by surface-adapted species of tropical army ants in each periodic nomadic phase of their activity cycle. The ants form this nest by hooking their bodies together hanging from a surface such as a log. The white objects are tens of thousands of larvae. Until these larvae mature, the colony remains in a "nomadic phase" of large daily raids and nightly emigrations, forming a new bivouac each night. Army ants are one of the few truly nomadic species of animals. (Photo of army ants on Barro Colorado Island by T. C. Schneirla. Courtesy of the American Museum of Natural History)

Growth of elk population is controlled only by predators and a limited food supply. Where protected, they accumulate in large numbers and must be fed to prevent starvation in the winter. Elk show a great deal of allelomimetic behavior, and this group of adult males move together as a unit. (Photographed at Jackson Hole, Wyoming, by J. W. Scott)

Group of young prairie dogs, *left*, occupies territory in center of prairie-dog town. When these young animals mature, the parents will move out to colonize a new area, thus avoiding overpopulation. All are feeding except the one standing upright, which is showing investigative behavior. They have been marked with dark patches of fur dye for identification. *Right*, recognition "kiss" of young prairie dogs meeting at mouth of burrow apparently enables them to recognize each other as members of same territory. Strange animals are driven out, preventing overcrowding. (Photos by J. A. King)

Effects of overbrowsing by elk herds in Wyoming. When food is scarce in the winter the elk eat all the branches of the spruce trees as high as they can reach. The lowest branches are protected by snow on the ground.

Allelomimetic behavior in an autumn migratory flock of blackbirds. Note that all birds are headed in the same direction and maintain a fairly regular distance from each other.

A colony of gulls on an island in Yellowstone Park, Wyoming. Like many other bird species, the gulls return each year to special localities for nesting. The nesting site is almost always an island, which gives protection from predators. Among the brightly colored adults can be seen groups of younger birds in the juvenile plumage. (Photo by J. W. Scott)

Adelie penguin colony at Cape Crozier, Antarctica. These highly social birds incubate their eggs and rear their young on land but secure their food from the ocean. Members of a colony such as this one must walk considerable distances to and from their home area. Birds removed from their home area can be used to test their methods of direction finding in a manner which is impossible with flying birds. (Photo courtesy of David H. Thompson)

Penguin chick chasing its parent for food. At this age the parent may run 50 meters before feeding the chick. The parent runs until only one chick is left, thus evading freeloading chicks belonging to neighboring parents. (Photo courtesy of David H. Thompson)

Rocky Mountains spend the summer at high altitudes where there is excellent browsing. At the first heavy snowfall in the autumn they start back down to their winter ranges, which may be ten to sixty miles away.

The vast majority of perching birds migrate from their breeding grounds, which are usually in a cold climate, to winter feeding grounds in warmer areas. The movements of salmon which migrate from rivers to ocean and back again in the course of their lives have already been mentioned. Even more extraordinary is the migration of eels. These were a favorite food fish caught in the rivers of Europe, and when the colonists came to America they found a very similar species in the North American rivers. No one had ever seen their breeding grounds. Meanwhile, zoologists were busy collecting and classifying the fauna of the world and, among other things, discovered a curious transparent fish in the Sargasso Sea. They named this new genus Leptocephalus. It was eventually discovered that these small fishes were really immature eels, and the whole remarkable story began to come out. Bred in the Sargasso Sea, the young eels gradually grow up and follow the Gulf Stream up through the North Atlantic. One species stops off in the rivers of North America and grows to maturity, while the other goes on across the ocean to Europe. The whole process takes several years, and the fish travel thousands of miles. It is possible that the young eels are simply carried along by the current and are attracted to fresh water at the proper time. The adult eels have to find their way back over a similar distance, and how they guide themselves is still a mystery.

Colonization. As well as migrating between two locations, animals sometimes leave an old locality and find a new one. On an occasional warm day in autumn the air seems to be filled with flying insects, wings glinting in the sunlight and bodies splattering against the windshields of automobiles. When we examine the bodies, they often turn out to be winged ants. The usual ant colony has no winged members, but at intervals broods of winged males and females are produced which then swarm forth all together. The males and females mate, and after a nuptial flight each pair settles down and attempts to form a new colony. A great many ants are killed or lost, but the behavior has the effect of spreading the species over the countryside and establishing new colonies wherever this is possible.

All animals have this problem of dispersal or distribution, and in some species there is a regular period in the life cycle which is devoted to the process. However, in many of the highly social vertebrate animals the processes of socialization to a particular group and localization to a particular spot strongly interfere with dispersal. In some there does not seem to be any regular and effective way of moving rapidly into vacant territory, and this creates a practical problem when a valuable species is threatened with extinction.

Nomadic wandering: The army ants. The vast majority of animals seem to have regular places to live. Even when they undertake enormous migrations like those of eels, it is between fixed locations. One of the few exceptions to the rule are the army ants which have been so thoroughly studied by Schneirla in the jungles of Barro Colorado Island in the Canal Zone. These creatures live only in tropical regions, and their unusual way of living has given rise to many legends about their ferocity and irresistible power. Although many of these tales are false, the facts are interesting enough without exaggeration. In the first place the ants do not destroy the vegetation as they advance, since army ants are entirely carnivorous. They eat chiefly insects and worms and, contrary to imaginative reports, never trouble any vertebrate which is big enough to get out of their way. They do destroy many fledgling birds, and a large animal which was completely crippled might be killed by multiple stings. The South American army ants would not be able to eat it, however, since they do not have the right kind of mouth parts. The African driver ants do have cutting mandibles and are able to eat larger animals caught in some kind of accident.

Ordinarily the ants have a central location in which a large number of individuals are grouped around the queen and the young. From this area the workers start out in one direction. At first there is a principal trail with a heavy traffic of ants upon it. As it gets farther away from the central bivouac, it spreads out like the branches of a tree. It is from the tips of these branches that the ants spread out to catch their prey. Ants may be going in both directions on the principal trails, and a veritable traffic jam sometimes results. Communication is maintained chiefly by scent trails and direct bodily contact. The raids usually start early in the morning, with activity lessening around noon. At the end of the day the ants return to the bivouac for the night.

Every so often the ants go into a migratory phase. Instead of coming back to the original site, all the ants, including those forming the bivouac, move out along the trail each night to a spot one hundred yards or so away. The new bivouac or nest is made again from the bodies of living ants, which usually find a protected spot under a log and begin to form a nest by hanging down from it in strings by small hooks on their legs. In this way a solid wall of ants is formed around the queen and young.

The wanderings of the ants are correlated with the reproductive cycle. As long as the young are in the pupal stage the ants stay in one spot, and when the new brood emerges they become more active and move from place to place. Weather conditions have some effect on the amount of movement and the location of the nest, but ants move on at the right time, whether or not the food supply in the locality has been exhausted. There is no indication that any one colony stays in a particular locality, and the army ants are thus a very unusual kind of animal society, one which is truly nomadic.

Limitations of movement. In addition to the behavioral processes which result in positive attraction to particular spots, there are other factors which limit the free movement of animals. Even army ants never move far in the course of a season because of the difficulty of moving while the young are being raised. Other species like clams and mussels move extremely slowly or not at all. Besides these internal limitations, there are all sorts of external barriers which tend to keep animals in one place.

Social barriers: Territoriality. We have already noted that most perching birds set up territories during the breeding season and attack any strange bird which crosses over the line. This effectively prevents free movement of the parent birds and keeps them close at home. Somewhat different arrangements are found in other animals. Most of the small rodents which have been studied live in definite locations or home ranges. They do not guard geographical boundaries, but, if they come into contact with other animals, fighting results, and they soon learn to avoid each other and stay fairly close to one spot.

We still do not know how widely territoriality may occur in the animal kingdom, since new cases are constantly coming to light. Many male fish, like the sticklebacks and sunfishes, defend small territories around their nests, and male lizards seem to defend particular areas. An interesting example of female defense of territory is found in a tiny Venezuelan frog. They are only about an inch in length, but the females station themselves along the edges of streams, and each one attacks any others that come within a foot or so. She first goes up to the intruder, exposes her bright-yellow throat and pulsates it rapidly. If the stranger does not retreat, the home frog jumps on it or wrestles with it until one is defeated and goes away.

Even some species of insects show territoriality. During the breeding season male dragonflies patrol an area about fifteen feet long and three feet wide along the edge of a pond and drive out any other males. As might be expected in such short-lived animals, this behavior lasts only about ten days for each individual, which is soon replaced by another.

Another example is the cicada killer wasp, whose territories are also maintained for about ten days. These are ground-nesting wasps that build their nests in burrows. Each male wasp sets up a territory enclosing some of the nest burrows, thus giving the male a chance to mate with any females that emerge from them. The territories are shaped like strips and may be as large as sixteen feet long and six feet wide.

As we have pointed out earlier, territoriality is by no means universal in the animal kingdom, occurring chiefly in vertebrates and arthropods, and somewhat sporadically in these higher animals, even in closely related forms. We conclude that territoriality is evolved only where it can be adaptive for the species. The conditions for adaptive function include the ability to recognize other species mates as individuals, a habitat which permits effective patrolling of territorial boundaries, and, finally, a daily activity cycle which permits surveillance of any intruder. Thus, some of the best examples of territories with precise boundaries are found among birds that are active in the daytime. From their aerial perches they are able to watch and patrol these boundaries effectively. On the other hand, nocturnal rodents can effectively defend only very small territories surrounded by larger overlapping home ranges. Among primates there are some small monkeys that guard definite territorial boundaries, but the common situation is one in which the troop has exclusive core area, usually including the sleeping place, with a larger overlapping home range. In man, territorial behavior largely reflects the culture in which a particular individual lives, including the way property is divided or shared among individuals, and the traditions concerning the importance of boundaries between societies. Ecology is also important, as there is no evidence of territoriality in the Arctic wastes inhabited by Eskimos. Thus, there may or may not be a biological basis for human territoriality, as there is no clear evolutionary trend. The implication that human beings are instinctively impelled to defend territories is unjustified. The current evidence for the origin of human beings on the arid plains of Africa indicates that any human group must have had to occupy a fairly large area in order to survive and that defending precise boundaries and maintaining surveillance against other wandering groups would have been impractical.

Many mammals mark particular spots by scratching or depositing urine or feces. While this may in some cases serve a territorial function, as it does in rabbits, this function cannot be assumed. In cheetahs the usual group is a family of a female and two or three younger individuals. As the group hunts, they regularly stop and mark spots with urine, much in the manner of dogs or wolves. When another cheetah group comes along, they follow the trail left by the first group for a short distance. Then, apparently having determined the direction of movement, they go off at right angles. This behavior is obviously adaptive in that it keeps the second group from hunting in an area from which all of the game has already been frightened, but equally obviously, it does not result in marking territorial boundaries.

Ecological barriers. In a great many cases the movement of animals is restricted by the kind of environment in which they can live successfully. The limpet cannot get too high out of the water or it is exposed to dryness and the attacks of land-living predators. Nor can it get too deep in the water or it will run out of the plant food on which it exists. Many land animals are kept in one place by physical barriers such as water or unfavorable areas like deserts. One interesting example is the butterfly Oeneis, which is found only on the tops of the White Mountains of New England, the Rockies of Colorado, and the coast of Labrador. For this butterfly a warm climate seems to be an effective barrier, and those in the different regions have been isolated ever since the last glacial period.

Biological barriers may be just as effective as physical ones. Penguins are expert swimmers and divers and can travel long distances over the ocean but are slow-moving and relatively helpless on land. They live only in Antarctica and certain isolated islands in the Southern Hemisphere, presumably because there are no carnivorous mammals like polar bears and foxes in these regions.

Locality and population. There are thus both positive and negative factors which tend to keep animals in particular places. One of the most important positive ones is primary localization, or the tendency to become attached to a particular spot in early life. Other types of behavior, such as shelter-seeking and nest-building, also keep an animal at home. In order to stay in a particular place, animals must have some power of orientation, and this is particularly important for animals which regularly migrate and have two homes. It is only rarely that animals show any real nomadic existence. Most of the reported cases turn out to be animals which are localized within a fairly wide area and follow a regular round within it. A wolf pack may live in a hunting range with a twenty-mile radius and follow a regular circuit or runway, reappearing every few weeks in the same spot.

There are also many sorts of limitations to movement. Poor powers of locomotion prevent an animal from going far. In many species there are social barriers to movement produced by a tendency to attack strangers and defend particular territories. Finally, there are all sorts of ecological barriers, including other species of animals, lack of available food, and physical obstacles. In fact, with so many external and internal factors controlling movement, the problem of colonizing a new area is a difficult one for many species.

All these factors which keep animals in particular places result in the formation of *populations*. A population is a major unit of ecological organization and is composed of the animals of the same species which live in a particular area. One of the most interesting things about a population is its numbers, and this in turn often depends upon its organization.

NATURAL POPULATIONS

The problem of numbers is an old one for students of both animal and human populations, but until recent years its study has been marked by theory and conjecture rather than observed facts. Everyone knows of Malthus's theory, which reasoned that since human beings have a natural tendency to increase their kind, and since the size of the world is limited, there should eventually be a point at which starvation, disease, and war will prevent any further increase in human numbers by killing off the surplus. Today we realize that the assumptions behind this theory are overly simple, but there is certainly a finite size to the globe and an ultimate limit to human population, barring the possibility of space travel. Animal populations confront this situation over and over again on a smaller scale, and it is interesting to find that in many cases there are definite means of controlling population size which are not dependent on starvation and competition.

Darwin's theory of natural selection was based in part on the same assumptions as that of Malthus and must also be modified to fit the facts which we have today. When we study animals in either natural populations or under laboratory conditions, we find that they are not collections of blindly reproducing individuals which are weeded out by the inexorable forces of nature but that they organize themselves into populations through their behavior and that this organization in many cases exercises a direct effect on population growth. Different species of animals show great differences in the degree of control over numbers. Understanding these facts has great practical importance for conservation and game management as well as far-reaching theoretical implications for human populations.

Fluctuating populations. There are certain species of animals which seem to act very much as Malthus would have predicted and which under certain conditions build themselves up to enormous numbers. There is usually a very high death rate for a short period, and the species almost disappears for a time, only to reappear again and gradually build up to the same peak of population. The various species of grasshoppers which inhabit arid plains regions are familiar examples. When the state of Kansas was first being settled, crops were frequently destroyed by swarms of grasshoppers. This does not occur so often since the western part of the state was put under the plow and the breeding conditions for the animals rendered less favorable, but overpopulation still occurs in other arid areas.

During the summer of 1937 there was a drought in the western part of the United States, and we made an automobile trip through the Dakotas and Saskatchewan. The wheat crop had completely failed in North Dakota, and nothing grew in the fields except Russian thistles about four inches high. Large grasshoppers without wings swarmed everywhere. They crawled over the highways, and where they were run over the survivors were eating the remains. We stopped to have a picnic and found the ground swarming with another species of small winged grasshoppers, which immediately covered the sweet-corn shucks we threw away and even hopped onto our sandwiches and attempted to eat the meat on them. Later on in Saskatchewan we saw a swarm of flying grasshoppers. The individuals were approximately a foot or two apart and filled the air as far up as we could see. It was a still day, the insects were not moving rapidly, and as we looked upward the air appeared silvery with their wings.

Such outbreaks as these appear to be associated with warm, dry conditions which are favorable to the survival of the young grasshoppers. The reproductive potential is very great, and when a large proportion of the eggs and young survive, swarms of grasshoppers appear. However, the same dry conditions are unfavorable to plant growth, the later death rate is high, and relatively few live to reproduce their kind.

Among mammals there are certain rodents which appear in similar hordes, and these have been extensively studied by Elton and his co-workers at the Bureau of Animal Populations in Oxford. The species which most commonly do this are the meadow mice Microtus (which are known in England as voles), ordinary house mice, and the lemmings of Scandinavia. It is difficult to determine the circumstances surrounding an outbreak, since it is seldom possible to spot in advance the locality in which it will occur. Hordes of mice sometimes appear where unthreshed grain has been stored in stacks, providing ideal food and shelter in the same place. Outbreaks of meadow mice and lemmings occur when there has been an unusual combination of climatic and food conditions favorable to survival. Usually it takes more than one such year to build up a big population of these rodents. Under such crowded conditions there is usually a period of high mortality, and the population sinks to a low level. Here again these species possess a high reproductive potential which if unchecked will produce populations larger than an area can support under average conditions.

In certain other mammals such fluctuations in numbers occur in a fairly regular way, and it has been possible to follow the history of one such species, the varying hare, over a considerable historical period. The Hudson's Bay Company of Canada has kept records of its fur catch of various animals since the late eighteenth century. Not only do the numbers of snowshoe rabbits fluctuate, but the foxes and other fur-bearing animals which depend upon them for food tend to increase and decrease in numbers correspondingly. The fluctuations are not absolutely regular in time but seem to run about ten or eleven years apart. As we see it in the field, there will be a good rabbit year in which rabbits are numerous everywhere. Then they almost disappear for a time, after which the numbers begin to build up again. There is nothing like the intense overcrowding seen in the grasshopper or locust populations, and considerable mystery surrounds the causes of the decrease in numbers. Following the peak year, there is a higher death rate than can be explained by the increased number of predators. One suggestion is that large numbers are favorable to the spread of parasites and diseases such as coccidiosis. On the other hand, there have been reports that dead animals found under these conditions do not appear to have been killed by any known disease. Their organs show deterioration of a type associated with extreme conditions of physiological stress. This has been called "shock disease," and it is possible that under overcrowded conditions the animals become so excited that they die or fall easy prey to infection. As a tentative hypothesis we may suppose that an animal which is either forced to leave its home locality because of lack of food or is overstimulated by contacts with strange animals becomes highly disturbed and frightened, and that if this is continued too long it may actually lead to death. The evidence from experiments with artificial populations supports the theory, since crowded mice develop enlarged adrenal glands typical of physiological stress.

We may contrast the varying hare of North America with the story of the European rabbit in Australia. This wild ancestor of our domestic rabbits was introduced into a country where it had no natural predators and promptly increased to enormous numbers. Few scientific studies have been made of the fluctuation of numbers, but it would appear that the rabbits rapidly reached a very high density of population and simply maintained it at that level. In Australia every year seems to be a "good rabbit year." Various attempts are made to control them, the most successful so far being the disease of myxomatosis introduced from South America, although the rabbits rapidly began to develop resistant strains. Whether the former high and stable level of population was due only to the lack of predators is not known. Like most of the animal species that have been domesticated, the European rabbit is highly social, and this may have some bearing on the lack of fluctuation of numbers, as will be seen below.

Other species in which overpopulation has been observed are the deer and elk. Persistent hunting in the United States has almost exterminated the wolves, coyotes, and mountain lions which prey on young deer. In most places hunting more than compensates for this lack, but where complete protection has been given, the results have sometimes been disastrous. In the Kaibab forest of Arizona the mule deer were protected for several years and gradually increased in numbers until they ate literally all the available food. Deer depend on browse, and they stripped the leaves and bark from the trees as high as they could reach standing on their hind legs. After this they died in large numbers.

In the early years of the national parks, when many species were threatened by extinction, the officials had a policy of trapping and killing predatory animals. The results of this policy and some other factors can easily be seen in the northern part of Yellowstone Park which was originally inhabited by a small elk herd. The numbers gradually built up but were kept somewhat under control by the fact that the herd migrated out of the park during the winter months into Montana, where it could be hunted. At one time it was the custom for hundreds of hunters to wait for the herd as it crossed the boundary on its usual trail and then to open fire. Whether because of fright or because more nonmigrating animals survived, the elk gradually ceased to migrate and thus spent the winter in Yellowstone Park, where there was only a limited amount of winter food. As the elk increased in numbers, the trees began to suffer the same fate as they had in the Kaibab forest, and the park service was faced with a serious problem of overgrazing. Such problems have arisen wherever predators have been eliminated and the herds completely protected from human hunting. Various ways of meeting these problems are being studied and put into effect, including the protection of certain predators. In Isle Royale National Park, in Lake Superior, a wolf pack has been protected for several years, and is apparently holding the moose population at a level where it no longer outruns the food supply.

Unlike the rodents, deer and elk do not have a high reproductive potential, and such build-ups in population occur over relatively long periods. Likewise it is more difficult to restore their populations once they have sunk to a low level. The problem appears to lie in the fact that these animals have a tendency to stay in a familiar part of the country in a relatively small range. Some, like the elk, have a seasonal migration, but this is very regular and each year the elk follow the same paths, which are presumably learned from generation to generation. Such a migration trail may be three to four feet wide and look almost as if it had been made by the hand of man. If numbers build up in a given territory through lack of predation, there is little tendency for the herd to move into a more favorable environment, and if in addition normal migration is blocked by fences or hunting, overpopulation appears to be inevitable.

In all these cases the species themselves have only limited control over their numbers, but even in these animals Malthus's theory does not quite fit the facts. When deer or elk live under conditions of partial starvation, their powers of reproduction decrease and fewer than the normal number of young are born. Thus there is some physiological mechanism which controls reproduction under unfavorable conditions and which often begins to act before actual extermination by starvation can take place. Predators normally limit the population before physiological control comes into play. The principal limiting factors are ecological in nature, and where conditions are sufficiently stable these factors tend to get into a sort of uneasy equilibrium which has been called "the balance of nature." There are, however, many species of animals in which these population build-ups do not occur and in which numbers are regulated by behavioral organization.

Stable populations. The social insects show some fluctuation of numbers with favorable and unfavorable environmental conditions, but they never appear in the enormous swarms seen in the grasshoppers. In a good ant year we can see an unusually large number of successful ant colonies, but their numbers never become excessive because they are regulated during the time of colonization. The vast majority of swarming pairs do not survive, and if they happen to land within the territory of an already established colony, they will be attacked and destroyed. This means that in a given area only a certain number of ant colonies can be established.

Even in the rodents there are species which do not show the tendency to overpopulation. Dr. John A. King has studied the prairie dog, one of the most highly social of ground squirrels. These animals formerly lived in large numbers on the western plains of the United States but always in well-organized colonies without any great surplus of numbers. Even in national parks, where predators are often reduced below a natural level and hunting is forbidden, there is no evidence of overcrowding. These animals have their colonies divided up into definite territories, each one of which is inhabited by a male and several breeding females. When a number of young have been raised, the adults in the territory leave and proceed to the edge of the colony, where they set up a new system of burrows in unoccupied territory, leaving the old burrows to the young animals. This provides for an orderly expansion or recolonization of empty territory with only a limited number of animals in any area. Their burrows and alarm calls protect them from predators, and under ordinary conditions they maintain large stable populations. In recent years their numbers have been greatly reduced by poison campaigns, on the theory that prairie dogs eat grass which would otherwise be available to domestic livestock. If they are ever allowed to reoccupy the grazing lands, it is obvious that recolonization will be a slow process. Favorable breeding areas are not continuous, and it would be difficult for prairie dogs to migrate over long distances.

When deer mice are raised in artificial populations with plenty of food and water but limited space, they usually quit reproducing at rather low levels of population density, presumably because of some behavioral or physiological control, although its exact nature is not yet known. Few females become pregnant, and the sizes of sex glands are reduced in both sexes.

An opportunity to observe woodchucks under similar conditions of plentiful food but under natural conditions with unlimited space was provided by an army ordnance depot in Pennsylvania, where the land was regularly mowed and maintained in the meadows that are the normal habitat of these animals. They turned out to be almost solitary. No woodchuck will tolerate another woodchuck in his close vicinity, except briefly during the mating season and while the females are rearing their young. The result is that large individual spaces are maintained around each animal and constant pressure maintained. When Frank Bronson and his colleagues attempted to trap all the woodchucks out of a small part of the area, they found that at the end of the experiment there were just as many woodchucks as in the beginning. Although the species is obviously successful, woodchucks pay a high price for their continuous agonistic behavior. On autopsy the animals showed unusually high rates of heart disease and enlargement of the adrenal glands.

While the populations of carnivores fluctuate with their food supply, they are almost never reported in excessive numbers. Among wolves, population limitation is probably achieved through social organization. The usual life of a well-organized wolf pack was described by Adolf Murie in his classical monograph "The Wolves of Mt. McKinley." A wolf pack consists of several adult animals of both sexes, and they spend most of the day lying around and sleeping. Late in the afternoon they assemble, howl in chorus, and go off to hunt, sometimes in groups but often spreading out. If one is successful in making a kill, he howls and the others join him.

The size of the pack and the mode of hunting depend somewhat on the prey animals. The wolves of Alaska that chiefly hunt caribou usually include four or five animals, while the Isle Royale pack studied by Mech, which lives off moose, includes about twenty animals, all of whom combine when an attack on a moose is made. This last pack has been studied over many years under the relatively constant conditions afforded by a national park on an island, and it has maintained itself at almost exactly the same numbers year after year.

Other wolf packs have been studied by Douglas Pimlott in the Algonquin Provincial Park in Ontario. Again, these animals are protected, but they never produce excessive numbers. Usually, not more than one female in a pack will produce a litter in a given year. Wolves spend almost all their time in the open, even in very severe weather, but as the pregnant female nears the time of parturition, she digs a den, usually by enlarging some sort of hole constructed by other animals. Other members of the pack stay with her and, after the young are born, continue to bring back pieces of meat and bury them in the vicinity of the den, with the result that the mother does not have to go off to hunt. Like dog puppies, wolf cubs are fed with vomited food beginning at about three weeks of age. All members of the pack share in the care of the young and the defense of the den against predators such as bears. The result is a relatively high survival rate for the young. While the number in the litter is only four or five on the average, wolves are long-lived animals and their reproductive potential is great.

Some hint as to the mechanisms of population control is given by the captive wolf pack at the Brookfield Zoo near Chicago studied by George Rabb. In this pack there are separate dominance orders in males and females, with the males being dominant over the females. All of the females come into estrus, but the most dominant female prevents them from mating, with the result that only one litter is usually born each year. It is possible that similar controls on mating are established in wild packs. On the other hand, wolf populations can recover rapidly if hunting and bounty programs are relaxed and a plentiful food supply exists, as has been the case in parts of Alaska. When this is done, a balance between the wolf populations and their prey is quickly established. Where wolves hunt caribou, they are chiefly successful in capturing newly born animals or adults that have become weak or diseased. The result is that the prey population never becomes excessive as it did in the case of the Kaibab deer.

The height of social control of populations, however, is seen in birds. Most passerine birds like the song sparrow have a strong tendency toward localization and the formation of territories. The average male song sparrow settles down less than a quarter-mile from his birthplace and sets up a territory. Some birds defend their territory the year around, while others migrate, but all react most vigorously to intruders during the early spring and summer. Each bird has a territory of about three-quarters of an acre from which all other males are excluded and in which all mating and nest-building takes place. This means that a given area always has about the same number of song-sparrow nests and produces about the same number of young each year. As adult birds die or get killed, their territory is occupied by the young. Any surplus is excluded and unable to mate or build nests. In addition, the mortality rate is high, especially among young birds. The level of population in any area is always the result of a number of factors.

Sea birds such as gulls and pelicans have nesting places on small islands which are free from most predators. These limited areas are divided into very small territories, so that only a limited number of birds can lay eggs each year, and the number tends to remain constant. In addition the number of eggs laid by these birds tends to be small, so that only one or two young are raised each year.

The sage grouse described in a previous chapter provide an example of a high degree of stable social organization which is centered around mating behavior and which consequently is directly connected with the problem of population. All mating behavior takes place on the breeding grounds, and there is evidence that the same grounds are retained generation after generation. In one case a highway ran through a mating ground, and in another a small airfield was located on it, but the birds kept trying to come back even though disturbed. After mating, the females scatter over wide areas. Since the semidesert country will not support large populations, an area of several acres may contain only a single nest. The limiting factor on population is probably the amount of available food, but in order for the species to survive they must have a mechanism whereby the thinly scattered birds are able to find each other and mate. This is provided by the breeding ground and its social organization. There appears to be no problem of overpopulation, and when undisturbed the birds maintain their numbers year after year. However, the populations are quite vulnerable if greatly reduced by excessive hunting or if the breeding grounds are disturbed. Once the grouse have been eliminated from a given area, there appears to be no natural mechanism for setting up new breeding areas. It is possible that some natural method of colonization exists, but this has not been described.

Stable populations which have some kind of mechanism for controlling numbers can maintain themselves at high levels with optimum conditions for the existence of the individual members. These population controls have apparently evolved under conditions of high density, since such species run into difficulties when their numbers are greatly reduced. For example, many of the sea birds lay only one or two eggs per year. The chances of survival are good, the birds live several years, and as long as there is a big population the species flourishes. However, if the population is greatly reduced, the reproductive rate is so low that it takes a long time to build up the numbers again, and there is a real danger that storms and other accidents will wipe it out entirely.

The system of population control by territory also has its dangers. It depends partly on the animals returning to or staying in the same locality. If a species is greatly depleted in numbers or eliminated from certain areas, recolonization tends to be slow. All of this presents a real problem when it comes to practical game management. Many of the game birds and mammals which originally inhabited North America were highly social and had stable populations. Their very sociality makes them vulnerable to hunting, and their natural systems of population control render them difficult to restore if their numbers are greatly reduced. The passenger pigeon, the buffalo, and others which have either become extinct or have been threatened with extinction are all highly social in nature, with low rates of reproduction.

The extinction of one species of social bird has been studied very carefully. This was the heath hen of Massachusetts, which was eventually confined to a very limited area on the island of Martha's Vineyard. Once the numbers sank below a certain point, they continued to fall inevitably in spite of every possible protection which could be given by human agency, and the last heath hen disappeared in 1932. One factor which may have contributed to extinction was the intensive inbreeding produced by a small population. Since these birds belong to the grouse family, all of which have a considerable degree of social organization connected with mating behavior, it is more likely that a minimum number of birds may have been necessary to maintain the social organization necessary for reproduction.

In contrast, species which have survived well against human competition are often forms with a high reproductive potential and a low degree of social organization, so that their populations come back readily from small numbers. A species like the house mouse may start up a new population in a vacant area from a single pregnant female, and the risk of complete extinction is almost nonexistent.

THE GROWTH OF POPULATIONS

Several attempts have been made to experiment with the growth of populations under laboratory conditions, using species of animals ranging from protozoans to rats. As this work goes on, it is apparent that it is still barely begun. Each species that has been investigated presents a new and fascinating series of problems, and their study has many practical as well as theoretical considerations for man's control of his biological environment.

The growth curve. The curve shown in figure 34 can be applied to the growth of almost anything, from the increase in numbers of a population of paramecia to the increase in size of an elephant, provided various irregularities and departures from a smooth curve are disregarded. Its form is deceptively simple. If the example given is considered to be a population curve, it can be interpreted in the following way. The population starts with one hundred paramecia, each of which divides in two in each generation, so that the increase in the population is doubled in each generation. After a turning point in the middle of the curve the increase in the population is halved in each generation.

It must not be assumed, however, that this provides a universal theory for population growth. The curve is actually two curves which have been joined together. The lower half is a general curve for unlimited growth, in which the amount of increase in the population is directly proportional to its size. The upper part of the curve is a general one for limited growth, where the amount of increase in the population is inversely proportional to its closeness to an upper limit.

The size of a population is actually related to its growth in many ways. Many of these involve behavior—sexual behavior, the care of the young, and agonistic behavior which limits the available breeding territory. In any natural population there are likely to be many limiting factors which show a complicated interplay with numerous growth-promoting factors. Without knowing these, we cannot predict the growth of a population from the shape of the curve in the early stages, a fact which is well to remember in

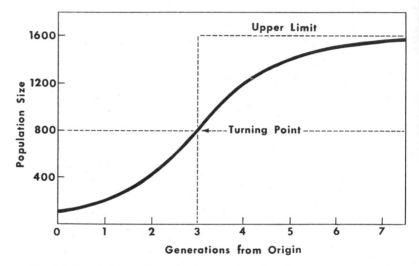

Fig. 34. An ideal growth curve. As is explained in the text, this is actually composed of two curves joined together, and the growth of many natural populations will fit it approximately. In the beginning stages, growth depends on the number of animals in the population; but in the later stages, large numbers are unfavorable to growth. The final shape of a curve cannot be predicted from its initial stages, since predicting the final size depends upon a knowledge of the limiting factors of growth. Also, in any natural situation the limiting factors fluctuate from time to time. (Original)

considering both human and animal populations. The growth of any population can be predicted only through a knowledge of the real growth-promoting and limiting factors. We are only beginning to appreciate the nature of the behavioral factors affecting population growth.

The limits of population. The late Raymond Pearl was much interested in population problems. One of his most interesting experiments made use of the fruit fly Drosophila, which is usually raised by geneticists in half-pint milk bottles on an artificial medium composed of corn meal, agar, and molasses. The agar is inoculated with yeast, and both the fruit flies and their larvae live upon it. In experimental work the amount of food and the space within the bottle are limited. Fruit flies are not highly social and are tolerant of each other in large numbers, so that we would expect that they would reproduce right up to the limits of the physical environment and then starve to death. However, when Pearl introduced different numbers of fruit flies into a set of bottles, he found that the reproductive rate, as measured by the number of eggs and larvae produced per female, was lower in the more crowded bottles. Two factors appeared to limit reproduction. One of these was the fact that the females did not have enough room on the surface of the agar to sit down and peaceably lay their eggs without being disturbed. The other was that they were undernourished and consequently less fertile. This is the same sort of physiological interference with fertility by limited nutrition as is found in the natural populations of deer, and it often shows up in experimental populations of other species.

Another insect which has been extensively studied is the flour beetle Tribolium, which occasionally gets into the flour bins of thrifty housewives and may increase and multiply to such an extent that the flour appears to be virtually alive. Professor Thomas Park of the University of Chicago has experimented with these animals by limiting the amount of space but giving them an unlimited food supply. If the flour is frequently changed, and the larvae sifted out and replaced each time, they survive just as well under crowded conditions as not. On the other hand, if they are allowed to remain in the same batch of flour, the crowded beetles begin to show increased mortality. The limiting factor of this population is the accumulation of waste products which render the food unfit to eat.

In these two examples we can see that while the limitation of space may have some effect, the populations are chiefly controlled by the limitation of food supply, which acts on the two species in different ways. These effects are not necessarily duplicated in animals with different behavioral habits, particularly those with a high degree of social organization.

During and after World War II, the city of Baltimore was the scene of an experimental project on rat populations. Always a problem in peacetime, rats become even more destructive of human food supplies during unsettled wartime conditions. Dr. John B. Calhoun was one of the many experimenters who took part in the project. He moved into the suburbs and set up a rat population in a limited area but with an unlimited food supply. The square pen was 100 feet on the side and contained 10,000 square feet of surface. A small group of wild rats from a known area were introduced, and the resulting population growth was studied. It is important to remember that these animals were a naturally associated population, because the organization or disorganization of a population may have a great effect on its growth.

If these rats had been raised in a laboratory, it would have been possible to rear 5,000 of them in comfortable cages having an equivalent total area. If unlimited space had been available and if the population growth had proceeded unchecked, the maximum potential growth of the population during the two-year period of study would have been 50,000. Actually the population reached a height of less than 200 adult individuals. Even with unlimited food the rats did not become overcrowded, nor were there any scenes of carnage. There appear to have been no outbreaks of serious disease, and predators were eliminated from the area. What actually happened was that the rats developed a considerable degree of social and ecological organization. A group of rats which grew up together tended to occupy a given area. They took food from the central supply and stored it away in more outlying holes. As more young grew up, the older animals established dominance over them, and the original group next to the source of the food supply tended to chase off any that came near. The rats in the outlying areas were able to obtain some food from the scattered portions, but only a limited amount. While they maintained themselves as adults, the reproductive rate in this outlying area was greatly decreased, apparently by some physiological means. The social organization in the outlying areas was less stable, and the young received less protection. Consequently there was a higher infant mortality, particularly from flies which lay their eggs on the fur of the young. While their territorial system is not as highly developed as in some birds, it is evident that the rats control their numbers chiefly by dividing up the space and keeping out strangers.

Disorganized populations. The Baltimore experimenters also designed an experiment in which there would be unlimited space but a limited food supply. This regularly occurs in many natural situations, and they converted such a natural population into an experimental one by the simple device of introducing a number of extra rats. The city of Baltimore is in many ways an ideal one for studying rat populations because of the custom of constructing houses in rows with back yards separated by board fences. When garbage or other food supplies are available, the yards make good places for supporting rats, and the regular design of yards and houses makes it easy to keep census records by trapping and releasing the rats. A census was taken of all the rats in one block. On September 23, it was estimated that there were 168 rats. On October 1, 75 of these were caught and marked and 10 days later 112 alien rats were released in the area. At this time it was estimated that there were 261 resident and alien rats present. During the following months the area was trapped again, and 37 marked residents and only 9 aliens were caught. At a capture rate of 50 percent, this indicated that only about 16 percent of the rats that were introduced in the population survived. When the adjacent blocks were trapped, only 2 alien rats were found outside the place into which they had been introduced.

Meanwhile the human residents of the block, who had not been taken into the confidence of the scientists, began telling each other how much worse the rats were getting. However, relief was soon to come, as the rats began to die off in great numbers. Rats were found run over in the street and lying dead in the alleys, and 21 were killed by residents and their dogs. The end result was that the total size of the rat population was reduced considerably below what it had been, with three times the mortality among the aliens as among the residents. In spite of the increased population pressure, few if any of the resident rats moved into other areas. They apparently have a strong tendency to stay in their home locality in spite of any disturbance, and the principal effect of the experiment was to decrease the total population. The high death rate was the direct effect of the disorganization produced by a sudden influx of strangers. After the disturbance, the population climbed to its former level in a few months, as a result of new births; it is difficult to disorganize a rat population permanently.

A classical example of social disorganization was inadvertently set up by the officials of the London Zoo, who wished to display a colony of baboons under seminatural conditions. In the late

1920s they trapped a quantity of wild hamadryas baboons, many of which were strangers to each other, and placed them together in a limited area from which they could not escape. At the start there were 100 baboons on "Monkey Hill," an enclosure only 100 feet long and 60 feet wide, smaller than Calhoun's experimental rat pen. There were many fights and a good deal of mortality from other causes, and two years later only 59 animals were left, mostly males. Then 30 strange females and 5 immature males were introduced. The result was continual carnage in which the males not only killed each other but literally pulled some of the females to pieces. Three years later there were only 39 males and 9 females, and the males were still fighting. In all this time only one infant baboon survived, so there was never any chance to build up a natural population. With our present knowledge we can see that the zoo should have started with a small related population and allowed it to organize itself naturally as it grew.

The sequel to the violence on Monkey Hill came some 30 years later when Hans Kummer studied the same species of baboons under natural conditions in Ethiopia. Unlike the savannah baboons described in an earlier chapter, and indeed unlike most nonhuman primates, the hamadryas baboons fight over the possession of females, and each troop is subdivided into small groups, each consisting of an adult male and three or four females with their offspring. Males will attack or threaten females that stray too far from them and thus keep the subunits separate from each other. The usual reaction is for the male to bite the female on the back of the neck. According to Kummer, he does this once a day on the average, but almost never inflicts a bleeding wound. The female does not fight back, but screams and crouches. Between males, fighting is almost entirely threats and bluff. In hundreds of encounters Kummer only saw two in which males actually touched each other in combat. When Kummer could determine the origin of fighting, it always arose over the possession of a female, but none of the fights revolved around a female in estrus. When a male died or was experimentally removed from the troop, the females rapidly joined other males with no indication that they were emotionally disturbed by their loss, as might be expected from the fact that contact is based on force rather than attraction. While these baboons show much more overt aggression and far less cooperation than the savannah baboons (the hamadryas only act as a larger unit when they assemble on a sleeping place in the rocky cliffs and when they leave it in the morning), they still do not show the bloody carnage that arose in the disorganized group in the London Zoo, where the males were held in constant contact and there were too few females to allow the formation of separate subgroups.

Such bad effects of social disorganization have many practical implications in wildlife management as well as for zoological parks. Bobwhite quails are naturally strongly attached to particular localities, and their survival through the autumn and winter months depends on the organization of coveys, which give mutual protection against cold and facilitate escape from predators. In past years it has been the practice to raise quails in hatcheries and liberate them among the wild population, in order to provide more birds for hunting. We would expect that the hatchery-raised birds would not be properly socialized, hence would not form natural social groups and, besides this, would survive poorly in a locality to which they were not attached. These expectations are thoroughly borne out by the facts. Census records taken before and after hatchery birds are released show that the final population is sometimes even lower than before the new birds were added. Like the rats, the strange birds not only perish themselves but upset the social organization of the native population and so cause a higher death rate among the native birds.

Most mouse species show a lower degree of social organization than the animals in the foregoing examples and a consequently greater tendency toward overpopulation, so that the results of disorganization are even more spectacular. Experiments with populations of wild house mice in limited areas show that they tolerate each other at a much higher population level than rats do if a plentiful food supply is available. If food and nesting places are close together, they will not move more than ten feet on the average. Since in a naturally formed population they become socialized to each other and do not fight, a large group can live in each small area. They only fight or avoid strangers from adjoining areas. This tends to produce a situation of pressure under high population levels which is somewhat analogous to the pressure maintained by gases. The molecules bounce off each other, and as a result the gas tends to push outward. As food supplies run low, the mice have to move farther and begin to fight with strangers. The chief difference is that while the gas molecules diffuse and move around through the containing space, each mouse in an organized population tends to remain in one spot. The situation is stable until food begins to run out and the mice are forced to move out of their accustomed ranges.

The effects of this pressure were observed by Calhoun while attempting to trap all the meadow mice in a given area. On the first days large numbers of mice were trapped and then, as expected, the numbers caught decreased rapidly. At this point the unexpected happened, and the trappers began to get more mice each day instead of fewer. It began to look as if mice were moving into the area from the outside as the resident population was trapped out. The same experiment was repeated with similar results in other localities where heavy populations were present. The experimenters concluded that under conditions of dense population and consequently high population pressure, a release of pressure at one point would cause the animals to migrate in that direction. If the migration proceeds inward from a roughly circular area, the animals coming in from the periphery of the circle will become more crowded as they approach the place where the pressure is being released. It is supposed that this is one way in which the mass migrations of lemming populations could get started.

Lemmings are small rodents common in Scandinavia and the northern part of Canada. There are old reports that the Norwegian lemmings regularly march to the sea and drown themselves, thus relieving population pressure. This sort of behavior seems almost incomprehensible to an animal behaviorist, since one of our basic theories is that behavior is adaptive and contributes to survival. Olavi Kalela and his associates at the University of Helsinki in Finland have studied the migrations of these animals in the field, and the facts turn out to be otherwise. There are historical records of mass migrations in which large numbers of animals have drowned, but the last of these seem to have occurred many years ago. It turns out that lemmings regularly migrate in the fall and spring, from a summer to a winter habitat, and back again. Although large numbers of animals are involved, they usually move as individuals. In northern Scandinavia the lemmings live during the summer in peatlands, breeding and increasing their numbers. In the fall they migrate into the forest to the alpine zone where the vegetation is heavy, the ground is dry, and there is a heavy snowfall. Under these conditions the lemmings should survive better than in wet peatbogs, which are flooded during the autumn and covered with a thick layer of ice in the winter. In the spring the lemmings migrate downward into the bog area again, where there is a better food supply during the summer. While moving, the lemmings do not appear panic-stricken but move rather deliberately. If attacked by predators while crossing open areas, they assume a defensive posture and are able to fight off predatory birds, at least. Nevertheless, there is a fairly high mortality during the migration. The lemmings do swim through lakes or streams that happen to lie in their path, and if these are wide, a considerable number may be drowned, but large numbers are successful in moving to the new habitat and surviving.

Large mass migrations have never been reported in the lemming species which inhabit North America, although there are regular and cyclic fluctuations in number. One species, the brown lemming, is reported to migrate and survive in its new habitat. The collared lemming lives on the tundra, and it is possible that the relatively flat habitat is too uniform to make migration adaptive, as it would be in the more mountainous regions of Scandinavia.

In general, the disorganization of a population results in a higher death rate for its members. Disorganization can be produced either by introducing large numbers of strange individuals into a population or by overcrowding due to natural population growth. The lemming and muskrat populations have poor internal systems of control, and this same low degree of organization makes them unusually susceptible to disorganized behavior under crowded conditions.

Undercrowding. While numerous scientists have demonstrated the bad effects of excessive population, the ecologist W. C. Allee emphasized other experiments which show that underpopulation is an equally serious problem. Protozoans and bacteria will not grow as well in extremely small populations as they will under more dense conditions, and this is also true of many higher animals. Goldfish living in groups are better able to withstand the effects of poisons than animals living singly, and many of the aggregations of animals seen in nature also provide mutual protection against unfavorable environmental conditions.

The problem of underpopulation is directly related to that of extinction. It is obvious that when a natural population falls below a certain level in a given area, the chances of its being wiped out by some sudden storm or environmental accident are very high. Added to this is the fact that a very small number of animals may be unable to give each other the mutual protection which is necessary for their survival. This is particularly true in populations of social animals in which a group of minimum size and special composition is necessary for the maintenance of social organization.

CONCLUSION

Our study of the effects of behavior has gone beyond social integration and communication to the new level of ecological organization. The process of socialization causes an individual to become attached to a particular social group. There is an analogous process of becoming attached to a particular locality which may be termed "localization." Negative factors such as social, biological, and physical barriers also tend to keep an animal in the same place and prevent free movement. The result is the formation of a group of animals organized on an ecological level. The unit of organization on this level is the population, consisting of the animals of a given species which live in a particular area. A population may in turn be made up of many subgroups organized on the social level. A primary characteristic of a population is its numbers, which depend upon the reproduction and survival of its members. Each of these in turn is strongly related to the adaptive behavior of the species involved.

When animal populations are studied empirically both in nature and in the laboratory, it is found that the original theory of Malthus, while it may have application in some species and under some conditions, is far too simple to explain all the facts. There are some species in which population size is limited chiefly by external factors, and they are often characterized by violent fluctuations in numbers. Even among these there are many cases where reproduction is physiologically suppressed under unfavorable conditions.

A great many internal and external factors affect the growth of populations, and one which has received considerable attention in recent years is that of social organization. The effect of social organization upon a population varies from species to species and according to different environmental situations. A high degree of social organization tends to protect the species against underpopulation by promoting survival. It may also protect against overpopulation through territoriality. On the other hand, a high degree of social organization tends to make a species vulnerable when there is a superior competing species, because of the numbers of animals massed together. Likewise, a highly organized population is likely to be rigid and be unable to adapt to changed environmental conditions, and it cannot recover so rapidly if the population is reduced to small numbers. The effect of social organization upon the population depends a great deal upon the nature of the species and upon the general environmental situation. The most general conclusion that can be made is that it is advantageous for the species to be flexible in its adaptations, both as individuals and as populations.

The relationship between social disorganization and destructively violent behavior resulting in death or serious injury has obvious implications for human behavior and provides a useful theory for the investigation of human violence. Such data as we now have indicate that crimes of violence are more likely to occur in areas of social disorganization, such as those produced by recent immigration and by crowding the newcomers into ghettolike conditions. We can further hypothesize that one of the causes of warfare is social disorganization on a vast scale, where the units are peoples and nations rather than individuals. On the other hand, the large-scale destructive violence occasioned by war itself is the direct result of social organization designed for the purpose of destruction. While animal societies tend to evolve agonistic behavior in forms which are adaptive rather than destructive, the very existence of agonistic behavior makes possible destructive behavior under conditions of social disorganization, or, as in the human case, organization of agonistic behavior for destructive purposes. The control of destructive violence is thus an organizational problem, and its solution lies in work directed along those lines.

These studies of nonhuman animal populations have obvious relevance for human problems of overpopulation. Are human populations like the unstable, fluctuating populations of certain rodents and thus (as Malthus thought) subject only to the ultimate limitations of starvation, predation, and disease, or are they more like other populations such as those of many birds, prairie dogs, and wolves, which are regulated by physiological, behavioral, and social factors? Historical evidence shows that human populations have always regulated their numbers in some way, through marriage systems, infanticide, and in modern times through various forms of birth control. Our current problems of overpopulation come chiefly from the relatively sudden reduction of disease as a limiting factor. We can predict that man, of all animals, has the capacity to establish effective social controls to meet this new situation.

The growth, limitation, and survival of populations is related to the behavior of its members through social and ecological organization. Among other things, this implies that in many cases the survival of populations is more important to a species than the survival of individuals, and this throws a new light on some of the older biological problems of evolution.

CHAPTER 11

Behavior and Evolution

A fawn which has been raised on a bottle often develops into a forlorn deer. Its only social relationships are with people, and as it grows older it shows the same patterns of behavior toward them as it might toward other deer. It follows them around, nuzzling for food, and if anyone tries to drive it off it rears on its hind legs and strikes out with its sharp pointed hoofs. In the fall it pushes people around with its antlers. It becomes a nuisance to the humans who fed it, but it will not go away, because like any other deer it has become attached to the locality where it was brought up. It cannot be turned loose in another place because it shows no fear of things that threaten deer under natural conditions. The only solution is to give it away to a zoo or wildlife park.

From the viewpoint of evolutionary history, this is a curious situation. Here we have an animal which cannot long survive unless it is reared in its proper social environment. This means that the same genetic factors which produce the animal have also produced similar animals which make up the kind of environment in which it lives. The environment has evolved along with the animal, and in the long course of evolution the two have developed in such a way that the animal can no longer adapt successfully, or even survive, when separated from its own kind.

The deer is not an isolated example. Birds reared by hand develop the same dependence on people and may be completely unable to mate with their own species even in captivity. As we have seen in the last chapter, the welfare of even the most lowly organized animals is partially dependent on their relationship to their social groups and populations. A solitary member of a complex social group like an ant colony is almost completely helpless.

All this means that evolution, by which we mean genetic change, is much more than a change in the power of individual adaptation to external "forces of nature." In modern terms, the individual animal must adapt itself to its social and ecological environments, which are in part a reflection of its own nature. In addition organized social groups and populations themselves develop new powers of adaptation at different levels of organization.

Importance of evolutionary studies of behavior. The evolution of behavior is bound to be an even more speculative subject than the evolution of form and structure, since the latter at least leaves traces of past conditions in the form of fossils. One may therefore raise the question of why this subject should be pursued at all, and if so, in what directions. There is obviously little point in repeating the collection of data corresponding to those of comparative anatomy, except perhaps in some disputed cases of relationships where additional evidence is needed, since the broad lines of evolutionary relationships have already been thoroughly established. Indeed, no broad general studies of behavioral evolution will be possible until behavioral evolutionists take a leaf from the book of the comparative anatomists and develop general systematic methods of describing and naming behavior patterns. At the present time most behavior patterns are described in colloquial terms, such as the "down-up" pattern of certain ducks. The name means something to those who know ducks but conveys little information to anyone else. Behavioral evolutionists still need to develop good conceptual tools. Further, the greatest payoff will come from concentrating on those evolutionary problems which are peculiarly behavioral in nature. That these are of fundamental importance follows from the position of adaptation as a central biological concept and the fact that for animals behavior is a major form of adaptation.

ADAPTATION AND ORGANIZATION

Individual adaptation. The behavioral adaptation of an individual depends on two sorts of abilities. One of these is its basic motor, sensory, and psychological capacities, which are largely defined and limited by its heredity. The second is its behavioral organization, somewhat defined by heredity but modified to a greater or lesser degree by previous experience. All animals show some tendency in this last direction, although it varies a great deal in complexity and duration.

A stentor has four or five patterns of behavior with which it can meet unfavorable conditions, and the effect of previous stimulation modifies its behavior for half a minute or so. We can easily predict what this single-celled animal can do on the basis of its structure and limited motor apparatus, and its behavior is therefore closely related to heredity. By contrast, a multicellular animal with a high degree of manipulative ability, like a raccoon or chimpanzee, can do hundreds of different things with its hands, and it is extremely difficult to predict in advance what it will do in any given situation. In even greater contrast, the effects of its previous experience may last several years and possibly over its entire life. Thus the higher animals inherit capacities to make a very wide variety of adaptations and combine these in all sorts of ways on the basis of the kinds of problems which they meet. This means that the specific type of adaptation made by a mature animal has a very distant relationship to heredity.

There seems to be a general evolutionary tendency for animals to develop a wide variety of adaptive behavior. The more the capacities for learning and for variable organization of behavior are present, the more it is possible for an animal to learn from its parents and pass the information along to the next generation. As we accumulate greater knowledge of natural animal behavior, we find more and more evidence that many animals possess the rudiments of this new ability, which we can call cultural inheritance. The migration trails of mountain sheep and the learned fears of wild birds are two of many examples.

Information can be transmitted from one generation to the next by either biological or cultural heredity. Usually there is some transmission of both types in any animal society, with interaction between the two, but there may be extremes in either direction. The life of the mallee fowl of Australia leaves almost no opportunity for cultural heredity. It is related to the gallinaceous birds like the domestic chicken, but is placed in a separate family with other similar birds that live in Australia and the East Indies. The species has evolved an unusual method of incubating the eggs. Instead of the female setting on the eggs, the male in the late winter season scoops out a hole in the sand, fills it with vegetation,

Fig. 35. Male mallee fowl in characteristic digging posture. These birds, which are about the same size as domestic chickens, build nests out of sand and organic matter, usually about 15 feet across and 2 or 3 feet high, and work constantly to maintain an even incubation temperature for the eggs buried in the mound. (Sketch of behavior described by H. J. Frith)

and covers it with a mound of sand. Decomposition heats up the sand, and the female comes to the mound, mates, and lays one very large egg approximately once a week. The male covers each egg with sand and visits the mound daily, uncovering the nest, and testing the temperature by thrusting its open beak into the sand. If it becomes too hot, the bird opens the nest early in the morning and scratches cool sand into it. As summer progresses, less heat radiates from the decaying vegetation, and the bird piles the sand deeper and deeper in order to provide insulation. Later in autumn, the ground begins to cool down, and the bird keeps the nest warm by opening it during the middle of the day and scratching in sand that has been warmed by the sun. In this way, the bird is able to keep the nest at a relatively constant temperature of 92° F throughout the long laying season.

Thus the bird is able to maintain a constant temperature in the nest through behavioral instead of physiological mechanisms, whereas most other birds maintain the incubation temperature through a combination of the two: the behavior of setting on the nest and supplying heat from the body metabolism.

In the process of incubation the mallee fowl may build a mound of sand fifteen feet across and three feet high, and move a large portion of it daily. Obviously, such behavior is enormously wasteful of energy, compared to the incubation process of other birds, and it could only have evolved in conditions where climatic fluctuations are small and predators are few.

The other remarkable feature of this behavior is the life history of the bird. The young chicks take eight weeks to hatch, compared with three weeks in the domestic chicken, and when they hatch they dig their way through two or three feet of sand and go off individually into the scrub where they pick up their own food from the very start. The newly hatched birds are so mature that they are able to fly on the first day. Thus, there is no possibility that information can be passed from one generation to the next by means of a learning process, except that the young bird could possibly remember the mound from which it came. Yet these birds grow up and repeat the whole process generation after generation.

This kind of evolutionary tendency in behavior can be contrasted with that in the higher mammals. The young sheep or the young monkey is part of a social group from the day of birth, thus having the opportunity to learn from its associates at all times, and there is considerable evidence that newly learned information is commonly transmitted in this way. Furthermore, behavior is not organized so that it is highly adapted to a particular mode of existence. A mammal usually has many simple behavior patterns that can be combined and recombined in a great variety of ways in order to meet changing environmental conditions. In the example of the sheepdog, the complex behavior of herding sheep does not appear in the absence of experience and training, but is based upon many different behavioral capacities which can be combined in this fashion. In an even more extreme case, it is difficult in man to find any instance of complex behavior which is not extensively organized through learning and experience, with the exception of a few very general motor patterns such as walking.

Once the ability to transmit cultural information is established, cultural traits have the possibility of evolving and changing in their own way, free from the limitations of biological heredity. The extreme example is human language, which changes and develops new varieties without any change in the basic heredity of the speakers. At the present time all our evidence indicates that cultural inheritance exists only in quite simple forms in animals other than man, but future research may show that it is more common and complex than we now suspect.

Social adaptation. One of the basic theories of animal behavior is that behavior is an attempt to adapt to changes in the environment. Almost all animals show some degree of social behavior, and this means that some of the changes to which animals adapt are found in the social environment. In fact, adaptation of this sort may sometimes take precedence over any other type.

When birds like robins or song sparrows develop bright plumage during the spring months and attract attention to themselves by sitting in conspicuous places and singing loudly, there is an obvious conflict in adaptive tendencies. From the viewpoint of individual survival, the behavior of the male bird is suicidal. Yet if the bird did not do this, he would be unable to obtain a territory and mate and so would have no descendants. The production and survival of offspring appears to be more important than the survival of the individual.

Another example is the erection of the white hair on the rumps of frightened pronghorned antelopes. Instead of running away and making themselves inconspicuous, the antelopes often circle the alarming object and seem to make themselves as visible as possible. This behavior apparently serves as a warning signal to other antelopes in the neighborhood, and the usual result is that several start running this way and that, in a manner which is probably confusing to a predator.

These traits which seem so useless or even harmful for individual survival actually permit the whole social group to adapt to general problems affecting the survival of all. The territorial display of birds solves the problems of overcrowding and of orderly recolonization of adjacent vacant territory. The alarm signals of antelope make it difficult for a predator to surprise any member of a group. The shrieks and dashes of parent robins may disturb a prowling cat so that he lets the fledgling escape. Even the energy wasting contests between buck deer in the rutting season as they attempt to round up the females may at least result in a male being near as each female comes into estrus, so that all have a good chance to mate.

Adaptation on the level of populations. Under conditions of food scarcity, the tendency of deer to become attached to particular localities is poorly adaptive for an individual and its local group. Yet the trait is very strong in these animals, and we can only conclude that the resulting stable distribution of whole populations is more important than local survival. We can state as a general principle that when there is a conflict in adaptation between two levels of organization, the higher one tends to take precedence. If an individual cannot survive without the survival of the population of which he is a part, individual adaptation must take second place.

This rule is more important in some animals than in others. Some species of birds have developed a stable social and ecological organization without which the individual is unable to exist. On the other hand, animals such as mice have developed a very unstable kind of social and ecological organization, survival being chiefly dependent on the ability of very small groups to live independently and so survive. The latter species are capable of withstanding great fluctuations in numbers.

These two kinds of adaptation, one stressing the survival of the social and ecological organization and the other that of the individual, often occur in species which are quite closely related from an anatomical point of view. Populations of mice are very unstable, whereas rats, which are much the same type of animal and once were included in the same genus, develop a more stable social organization. However, the two trends are not mutually exclusive. A high degree of social organization may also be accompanied by considerable adaptability of the whole population under changed environmental conditions. Rats develop considerable social organization, but if this is disturbed it does not necessarily cause the death of all its members. Another example is that of the starlings which were able to set up large populations in a different environment when brought to the United States. The same thing goes for most highly social animals which can be easily introduced into new environments. Elk appear to live equally well in zoos, on the plains, or in a forest environment. Other populations, like those of sage grouse and pronghorned antelope, have much more limited adaptability.

All this leads to the conclusion that adaptation is much more than a matter of individual adjustment to changing conditions. The more we understand about individual animal behavior, the more important social and ecological organization seems to be. These discoveries are paralleled by a similar set of new facts from the science of genetics.

POPULATION GENETICS

It is only recently that geneticists have begun to realize that one of the fundamental techniques of their science is the analysis of inheritance in relation to populations. Mendel discovered the ratios which bear his name because he raised a population of peas and took a census of the different types. By use of this technique his followers eventually established the general principle that the vast majority of plants and animals pass along biological heredity through chromosomes. From this they went on to establish the theory of the gene and many other principles of inheritance. Eventually they began to wonder if these laws derived from the highly controlled and artificial populations of the laboratory would also apply to natural populations.

Hardy's law: The constancy of gene frequency. Oddly enough, the first approach to the genetics of natural populations was purely theoretical. Knowing the chromosomal mechanism of inheritance, we can predict that under perfect conditions the genetic factors present in the parents will be passed along to the offspring in the same relative numbers, so that there is no change in proportion. A parent having the pair of genes A and a has an equal chance of passing either one or the other along to each offspring, and if it has a large number of young, they will possess the two genes in the same 50-50 ratio as the parent. The same principle can be extended to populations. Suppose that we start out with a population with 0.1 aa individuals and 0.9 AA, in which there are 9 times as many A genes as a. If the adults breed at random and have infinitely large numbers of offspring, the genes will become distributed in the combinations of aa, Aa, and AA in the ratio 0.01 to 0.18 to 0.81, but the total number of genes in the population will remain in the proportion of 0.1 a to 0.9 A. Thus the chromosomal mechanism has the effect of keeping genes present in the same proportions but continually trying them out in new combinations.

In actual fact, real populations almost never meet these ideal theoretical conditions. They are seldom infinitely large and almost never so disorganized that real random breeding takes place. This means that in natural populations genes are not passed along in the same proportions as they existed in the parents, so that in the example given above the proportion of *A* genes might change from 0.9 in the parents to 0.8 in the offspring. In this way heredity changes from generation to generation. This is a basic principle of evolution: Limitation and organization lead to genetic change in populations.

From the viewpoint of genetics, the problem of evolution becomes a matter of determining the factors which cause changes in the proportions of various genes present in a population. The geneticist Sewall Wright has studied these factors in great detail and concludes that they can be grouped under the following headings.

Mutation pressure. Mutations, or direct changes in the genes, occur in nature as a result of various sorts of chance disturbances. They usually occur only rarely, but over a long period of time it would be theoretically possible for all of a given type of gene to mutate and so transform a population. In many cases there is a tendency toward reverse mutation, so that a balance between two genes is the eventual result.

Selection pressure. Second, there is the factor of selection produced by differential rates of survival and reproduction. Its action depends upon the severity of selection and the kind of inheritance in question. Selection produces a rapid change in a population if a trait is controlled by one gene and is not much affected by chance environmental factors. The change is much slower if the trait is produced by a combination of genes or is strongly affected by the environment, as is so often the case with behavior traits.

Inbreeding and genetic drift. Finally, inbreeding has the effect of taking genes out of the heterozygous condition (*Aa*) and putting them into homozygous combinations (*AA* or *aa*). In a large population this will not alter the proportions of various genes, but if the population is cut down to small numbers, continued inbreeding will result in the accidental selection of one gene in one inbred line and another in others. Therefore, the division of a population into small subpopulations, or even its reduction in numbers to a low level from time to time, will cause the proportion of genes to drift at random. This means that such populations continually change and become different from each other.

These principles lead to the idea that populations may be more important units in evolution than individuals, and that the organization of a population, that is, whether it is divided into small or large subgroups and whether or not mating is done on a basis of random assortment, will determine its evolutionary change. This immediately relates evolution to populations, social organization, and ultimately to animal behavior. There is not enough space here to summarize all the complexities of the current genetic theories of evolution, but in general the most favorable conditions for evolutionary change are those in which a species is divided up into a large number of moderately small populations. Whether this happens or not often depends upon the behavior of the species concerned.

THE ORIGIN OF BEHAVIOR

Modification of preexisting patterns. The problem of how new behavior patterns come into existence is an interesting theoretical problem. It is also a largely speculative one, since we rarely have any opportunity to study behavior patterns as they first appear. One obvious possibility is that new behavior patterns may arise as modifications of preexisting behavior, as J. S. Huxley found in his studies of the mating behavior of the grebe. Behavior once directly adaptive may be reduced to having a social signaling function, and he called this process *ritualization.* We must remember, however, that the reduction of behavior to social signaling occurs developmentally in the lifetime of an animal as well as in its evolutionary history. For example, agonistic behavior in a flock of chickens is reduced to dominance and subordination signals as a result of the fighting experience of each individual.

Preadaptation. If all behavior came from preexisting behavior, there could be only a limited number of behavior patterns. As one kind of behavior took over a new function, its old function would be lost. Obviously there must be ways in which new patterns may emerge. One possibility is simply that new behavioral capacities may occur accidentally, partly as a result of mutation and the resulting genetic variation, and partly as the products of evolutionary changes of other sorts. This is essentially the theory of *preadaptation,* and suggests that complex animals have a much greater variety of behavioral capacities than are ordinarily used.

A few years ago the blue tit, a wild bird in Great Britain, began to open milk bottles as they were left on the customer's doorsteps. This was a new behavior pattern, as milk bottles had never before been available to the species, but the birds obviously had the adaptive capacity to make use of this new source of food. The fact that animals do not use all the capacities available to them is related to the general principle of conservation of effort (whether nervous or muscular). Animals usually adapt by using those patterns of behavior requiring the least effort, with the result that many capacities are not exhibited in normal day-to-day living.

Modification of learned adaptation. Still another possibility is that animals may have general adaptive capacities whereby problems may be solved by a process of trial and error and subsequent habit formation. Where the same problems occur generation after generation and their solution is of major importance for the species, there may be selection in favor of animals that solve the problems more quickly, so that eventually very little trial and error and learning is involved. This hypothesis is known as the "Baldwin effect." For example, it is hard to see how the woodpecker finches of the Galapagos Islands described in chapter 7 could have developed their complex patterns of feeding by accident. Unlike woodpeckers, finches have no strong beaks that could enlarge the hole of a wood-boring insect. Instead, the finch takes a thorn in its beak, inserts it in a hole, levers out the insect larva, and eats it. Some frustrated ancestral finch may have solved this problem by a process of trial and error, and subsequently there would be selection for finches that performed the act with less and less learning.

THE EVOLUTION OF BEHAVIOR

Active and passive adaptation. A key concept of Darwin's theory of natural selection is that of adaptation. The animal which is better adapted to its environment is more likely to survive. Adaptation can be defined as a response or adjustment to a change in the environment. Seen in this light, behavior *is* adaptation; and among animals that show behavior, it is one of the major ways in which adaptation is accomplished. For those animals which do not show behavior, such as sponges, adaptation can occur only through changes in physiology and structure. We can therefore classify adaptation as being either active, including behavioral and physiological adjustments which occur as immediate responses to change, or passive, including changes in structure which can occur only from one generation to the next. An example of passive adaptation is the change toward darker pigmentation in the moths of the "black country" of England, as the vegetation and tree trunks on which the moths rested became darker as the result of industrial air pollution. An example of active behavioral adaptation, on the other hand, is the evasive action of a flying moth as a predator attempts to catch it.

In either kind of adaptation, variation is essential. In the case of form, variation is chiefly produced by gene mutation, whereas in behavioral adaptation, variation is produced by changes in the behavior itself. Consequently, it is important to measure variation in behavior as one of its fundamental properties. The early students of animal behavior were typologists and assumed that behavior was fixed and invariable, but this is by no means the case. Even such stereotyped behavior as the phototropic reaction of Drosophila shows a great deal of variation when it is actually measured and timed.

Species adaptation vs. individual adaptation. So far in this book we have talked of the adaptive responses of individuals, both singly and in populations. The term adaptation is also used in another sense: to mean changes on an evolutionary time scale in a species—not an individual—in every sort of adaptive capacity, whether active or passive. Thus there are two general processes of adaptive change, one of the species as a whole, which is largely based on genetic change over generations, and one of individuals and social groups, based on behavioral and physiological change. The primary phenomenon, however, takes place in the individual, since the success or failure of its adaptation leads to selection and thus determines the direction of evolutionary change.

Fixed vs. flexible patterns of behavior. There are two general ways in which the behavioral variation necessary for adaptation can be met. The first is the evolution of a number of relatively stereotyped patterns of behavior that can be used alternately and attempted in succession (the so-called "fixed action patterns"). Such an arrangement is characteristic of the lower animals, such as the stentor, whose four or five different reaction patterns to sediment were described in an early chapter of this book.

The second general way of meeting the need for behavioral variation is the evolution of a few simple and flexible patterns of behavior which can be extensively modified by learning and experience. This is the direction in which many mammals have evolved. Man has only a few fixed action patterns, such as the social smile, and shows an extreme example of a flexible behavior pattern, that concerned with language. Starting with the pattern of making a variety of babbling noises and the capacity to make sounds voluntarily, the human infant is able to transform these into a vocabulary of thousands of words and use them in an almost infinite number of combinations. Furthermore, within a certain critical period of development, he is able to easily learn a variety of different languages, multiplying his capacities still further.

However, in no case is behavior created solely by experience. There must be some sort of initial organization of behavior on which the process of learning and the further organization of behavior in this fashion can act. Finally, these two ways of producing behavioral variation are not necessarily exclusive and, particularly in the higher animals, one can find examples of both in the same species.

It is obvious that a highly organized fixed pattern of behavior is useful only where the same situation requiring behavioral adaptation occurs over and over again and generation after generation. In short, this kind of evolution will result where there is a stable environment with a relatively small number of changes that regularly reoccur. Such a stable physical environment is sometimes found in the ocean, which is buffered against changes in temperature and physical composition.

Another stable environment is that provided by a host for its parasites, and this accounts for the reduction of behavioral capacities seen in parasites as compared to their free-living relatives. Internal parasites inhabit an environment rendered highly stable by homeostasis, and even the external parasites adapt themselves to conditions that are almost identical generation after generation. In this respect, the machine-like behavior of a tick, which after its early development climbs to the top of a bush or shrub and waits until stimulated by a passing mammal, is an example of behavior that has little more variation than a physiological reaction and shows almost no modification of the result of previous experience.

For land animals, and particularly those in the temperate and arctic zones, the situation is quite different. Here, the most stable part of the environment is the social environment, and this is where examples of complex stereotyped behavior are likely to occur. For a bird, the social environment remains the same for generation after generation, and it is likely to develop a number of fixed patterns of social behavior, whereas its food source varies from day to day and from season to season, and its ingestive behavior may be quite flexible.

In any case, there is no way of distinguishing a final form of behavior which has been organized in a stereotyped way through a process of growth from another which has been stereotyped by habit formation. The fact that behavior is relatively invariable gives no clue as to how this condition has been produced. The development of behavior must be experimentally analyzed before firm conclusions regarding the nature of its organization can be drawn.

Evolution of social behavior: Aggregations. Because the social environment forms the most stable part of the environment for a land-living animal, and because it becomes increasingly important as social organization is developed, a consideration of the evolution of social behavior is highly important to any theory of behavioral evolution. Of the nine major behavioral systems, four are found almost universally in any animals capable of behavior. These are: ingestive, shelter-seeking, investigative, and sexual. Only the last is primarily social in nature, and for many animals living in the sea, sexual behavior patterns are unnecessary, the germ cells simply being released into the surrounding water. The almost universal occurrence of sexual reproduction, on the other hand, seems to be correlated with the fact that by providing a mechanism for genetic variation, it becomes essential for rapid evolution. While there are some exceptional species that reproduce asexually, it is the sexually reproducing species that have tended to survive.

The late W. C. Allee was interested in measuring the physiological benefits of social behavior and social organization, and concluded on the basis of a large number of experiments on animals, ranging from paramecia to goldfish, that animals survived better in groups than they did individually. In many situations the body of an animal of the same species provides a more favorable environment than the rest of the surroundings. Consequently, the animals' behavior in seeking such a favorable environment (shelter-seeking behavior) will account for the initial formation of social groups. Indeed, the vast majority of social groups in the lower vertebrates are temporary aggregations formed in this way, although there are also many instances of temporary aggregations formed through sexual behavior. In any case, once a group of animals has been brought together, there is an opportunity for evolution of higher forms of social behavior.

Evolution of social behavior: Higher forms. The evolution of more complex forms of social behavior is dependent upon the capacity of one individual to discriminate between others. The most primitive capacity of this sort is the ability to discriminate between members of the opposite sex, and in many of the species of arthropods and vertebrates, males have conspicuously different markings from females. In the fruit fly, *Drosophila melanogaster,* the male has a conspicuous black band on the tip of the abdomen, while the female has a series of narrow bands. Another primitive form of recognition is that differentiating between young and adults. Besides the differences in size, the adults may be conspicuously marked in different ways as they are in many species of birds, the adult markings not appearing until the beginning of the first breeding season. Such differences facilitate the evolution of parental care and the two related systems of care-giving and care-soliciting behavior.

Sex and age differences allow discrimination between large classes or castes of individuals, but most highly social animals go a step further and are able to differentiate between similar individuals in a group. In a flock of domestic hens, each individual is able to recognize every other individual in a flock of at least 100 hens.

The evolution of higher forms of social behavior and the resulting more complex social organization is thus dependent upon the development of three general capacities. The first of these is the sensory capacity to rapidly discriminate between individuals. This can be achieved through a variety of sense organs, chiefly sight in most birds, and often odor in many mammals, especially those that are nocturnal in their habits. The second general capacity is the existence of genetic variation in the species so that each individual has a unique identity. Song birds vary their songs, and neighboring birds learn to discriminate between individuals. Even in birds, such differences are augmented by differences resulting from differential early experience and environmental situations.

A great many variations in appearance are selected against because they make the animal less viable or more conspicuous to predators. Therefore, the variations that persist tend to affect nonessential characteristics. In domestic chickens, birds recognize each other chiefly by differences in the size and shape of the comb and wattles. Most wild rodents of any given species are very similar in external appearance, but experimental evidence shows that they can recognize individual differences in odor. The genetic mechanism underlying such differences will be interesting to study as better methods for recognizing and analyzing odors are developed. Further, the more successful a social species is in protecting its members, the greater the degree of variation that it can tolerate. Consequently, polymorphism should be common in an ecologically dominant social species, as is the case in wolves.

A third general capacity is that for learning and remembering the discrimination between other individuals of the same social group. Such capacities are highly developed in many social vertebrates, even in species like birds, whose behavior is often considered to be highly stereotyped. On the other hand, discriminations among arthropods are of a much simpler sort. In the social insects there are caste or group differences of a conspicuous nature which could be utilized by a fixed action pattern, and most of the learned recognition appears to be based on a general colony odor.

Agonistic behavior. As with other forms of complex social behavior, the evolution of agonistic behavior is dependent upon the ability to discriminate between individuals. Indiscriminate attacks on members of the same species would be maladaptive under most circumstances; and hence selected against. Agonistic behavior or social fighting is, therefore, confined to the arthropods and vertebrates.

Agonistic behavior has been independently evolved in a large number of animals. Assuming that it is more likely to have developed from preexisting behavior than by a sudden mutation, the most probable origin of agonistic behavior is through modification of defensive behavior, which is found almost universally in all animals that are capable of movement. An animal attacked by a predator will attempt to strike back or escape, depending on its motor capacities. A large number of animals have behavior similar to that of a cornered rat or mouse, which simply turns and bites anything that approaches it. This almost reflex behavior can also be adaptive against accidental injury by a species mate. Once such behavior has appeared in a social context, it could evolve in a great variety of ways. It is not necessary to assume that all agonistic behavior arose in this way, but we can definitely say that it is unlikely that agonistic behavior had a common origin in predation. In the first place, a great many animals that are strictly herbivorous show agonistic behavior; and in the second, the predatory animals develop agonistic behavior patterns which are distinctly different from those used in predation. The weapons, such as teeth or claws, that are used in predation are available for social fighting, but are used in quite different ways and ordinarily in ones which are much less harmful.

Social fighting has a variety of functions, not all of which may be found in the same species. The most general effect of fighting is to cause an injured animal to move away from another, with a resulting regulation in the use of space. Sometimes this regulation is associated with particular areas of ground, with the result that territoriality develops; in others, such as the large herd mammals, the behavior simply keeps individuals spaced out a few feet away from each other, no matter what the area may be, and the result is the preservation of individual space. In wolves, which are social carnivores, agonistic behavior results in the division of food and also regulates those animals that have access to each other for mating. The same species also shows a defense of small territories around the den. In domestic dogs there are even variations in function between breeds. Basenjis are tolerant of each other with respect to space, but develop a definite dominance order over food, while Shetland sheepdogs tend to share food with little difficulty, but are intolerant of animals in the same space.

In general, the functions of agonistic behavior tend to evolve in ways which promote the survival of the species. Thus, territoriality in birds assists in providing favorable conditions for reproduction and the survival of the young. Spacing resulting from agonistic behavior produces a more efficient utilization of food supplies, and thus safeguards against accidental destruction of a group. On the other hand, agonistic behavior may also evolve in ways which only benefit the survival of the genes of a particular individual in competition with others. The butting contests between male mountain sheep during the breeding season directly benefit only the particular males that win, although it has been argued that this also results in selection for animals of greater physical strength. Such behavior also favors the selection of individuals with heavy horns, which are of no value in escaping from predators, and even may be a considerable handicap.

In spite of the fact that wild sheep are highly allelomimetic and show concerted behavior in most aspects of their lives, these fights are always between individuals. One never sees a group of males attacking another, nor do the fights ordinarily end in fatal injuries. This is an excellent example of ritualization in the sense of reduction of behavior to a largely symbolic form. The mountain sheep carry this even further in the development of dominance orders outside the breeding season. In the process of developing dominance, fighting is reduced to signals, rather than overt attempts at mutual injury. In short, ritualization may be a developmental, as well as an evolutionary process.

The function of agonistic behavior in the so-called solitary animals has been little studied, except for a few cases like that of the woodchuck. In contrast to group-living rodents, like the prairie dog, where fighting is highly controlled, agonistic behavior in woodchucks is frequent and more serious. Agonistic behavior in the solitary carnivorous mammals is just beginning to be studied. One interesting example is the fox, whose normal group is a mated pair, the young living together only while they are still dependent. When placed in groups, foxes show all of the agonistic behavior patterns of dogs and wolves, except those connected with dominance and subordination. Such species have no need for this kind of behavior.

Animal species obviously vary a great deal with respect to the functions and the importance and amount of social fighting exhibited. Few actual figures are available, but the study of Phyllis Jay on Indian langur monkeys shows that an undisturbed and well-organized group spends most of its time in activities unrelated to agonistic behavior. In general, this is typical of well-organized animal societies. The occasional examples of destructive violence are the result of social disorganization, resulting either from human interference or environmental accidents. For example, house mice normally show agonistic behavior, which results in their spacing themselves out from each other. On meeting, two male mice usually show mutual avoidance. It is only when they are held together in a small pen where the beaten individual cannot escape that they go on to produce serious injuries and death. Agonistic behavior evolves to perform a useful function or functions within a definite kind of social organization. If this organization is destroyed, agonistic behavior frequently becomes disfunctional.

Evolution of allelomimetic behavior. This kind of behavior, in which individuals closely follow and mimic each other's activities, can only evolve in species that have sense organs that are well enough developed so that continuous contact can be maintained. The species must also live in a habitat which does not interfere with contact. The compound eyes of insects apparently are not sufficient to provide the precise visual information necessary for mutual coordination, and one of the few good examples of allelomimetic behavior in arthropods occurs in the army ant, whose colony members literally keep in touch and so are able to follow each other through tactile and olfactory senses. In other invertebrates it probably occurs in squids, which have well-developed eyes and often live in groups in the open water. However, allelomimetic behavior is primarily found in vertebrates, in those species that are diurnal, and usually those which live in the air, in open water, or on the open plains.

The functions of this behavior are primarily those of providing safety, which suggests that it may have its origin in the mutual protection afforded by aggregations. Allelomimetic behavior produces mobile and permanent aggregations. Many species of fishes live in schools, and these are particularly common in young fish. The school may break up later as the adults take up breeding territories or begin to reside in particular spots. Some ocean fish, like the herring and mackerel, live in schools all of their lives. Fish keep in contact with each other through their eyes, but also, to some extent, through sound and perception of water currents, as is necessary at night. Moving together as a group may have some function in finding food (minnows move along together as they feed), but its principal function can be demonstrated by simulating an attack by a predator. In response, the members of the school swim violently in all directions, sometimes stirring up mud from the bottom, and, in other cases, stirring up water so that vision is difficult. At the very least, the predator has difficulty in keeping his eye on any one individual.

In birds, allelomimetic behavior is the rule rather than the exception, but it may be limited to particular seasons of the year, as it is in the red-winged blackbird. Its principal function is that of providing safety from predators, partly because the flock has dozens or hundreds of eyes, and partly because if one bird reacts to danger he is seen by all and the whole flock is warned. In some cases there is probably facilitation of food-getting, as when a flock of gulls follows a fishing boat. When one gull tries to get a bite of food, his movement is observed by all the rest, who immediately come over and also look for it. Allelomimetic behavior has a special function in certain songbirds, which attack a potential predator, such as a hawk or an owl, in a concerted fashion. This is what the English ornithologists call "mobbing."

Among mammals, allelomimetic behavior is very rare in rodents, which almost never move in flocks or herds. Even when artificially crowded together, their movements are not coordinated. On the other hand, it is a major system of behavior among the large hoofed mammals, most of which are diurnal, and many of which live on open plains, tundras, or deserts. A typical example is a flock of sheep, whose members spread out grazing, coordinating their movements only loosely. One becomes alarmed by a predator or similar threat and runs toward his neighbors, after which all react in this same way, forming a solid bunch. Then the leader of the flock takes off in another direction, followed closely by the others.

In the pack-hunting carnivores, as in several species of the family Canidae, allelomimetic behavior has another function, that of cooperative hunting against large prey animals such as moose. Locating the prey does not require such coordinated behavior, and the pack may separate in order to do this. Wolves also defend their dens as a group against larger predators such as bears.

Finally, allelomimetic behavior is highly developed among most primate groups, particularly those that live on open plains or in large groups in the treetops. The behavior again seems to have the principal function of providing warning against predators, although combined defensive behavior is also seen in troops of baboons.

In summary, allelomimetic behavior has evolved in many species of animals, but still maintains its broad general function of providing security. We can also see in this behavior the beginnings of cooperative behavior, in which individuals combine their efforts to produce effects which are impossible for an individual. This behavior is, of course, immensely amplified in some human societies.

THE EFFECT OF BEHAVIOR ON EVOLUTION

Behavior, isolation, and inbreeding. The effect and importance of isolation on evolution has been recognized ever since Alfred Russell Wallace made his studies of island fauna over a century ago. We now realize that isolation is important because it tends to reduce the size of populations and produce inbreeding. Isolation is in turn the result of all the various factors which keep animals in one locality. Behavior is important both in producing a positive attachment to a locality and in providing territorial barriers against free movement.

But what is the explanation of cases in which two species or subspecies look almost alike and even inhabit the same locality, but one will not mate with the other? Two subspecies of the deermouse *Peromyscus maniculatus* have ranges which partially overlap, but they do not ordinarily cross-breed. Apparently the original species changed as it spread around the Great Lakes, and when its descendants met and shared the same range again, they were so changed in behavior that they no longer could amalgamate. Since the two subspecies, *gracilis* and *bairdii*, mate readily in laboratory cages, we wonder what behavior prevents this in nature. The explanation lies in the fact that one of these subspecies normally lives in woods, while the other occupies open grassland. When laboratory-raised animals of the two kinds were set free in a room with an artificial prairie and woodlot to choose from, each set up housekeeping in the environment of its ances-

Fig. 36. Habitat selection in two subspecies of the deermouse. *Peromyscus gracilis* (*A* and *B*, long ears and tail) lives in forests and brushy areas, while the subspecies *maniculatus bairdii* (*C* and *D*) normally lives in open fields and meadows. Harris took the descendants of laboratory-raised deermice of both kinds and tested each animal individually with regard to its choice of two habitats. One habitat contained tree trunks with flat plates attached to the top to give a semblance of shade and protection, while the other contained artificial grass made of stiff manila paper. In spite of never having had direct experience with the natural environment, the two subspecies preferred the artificial environment which was closest to that of their ancestors. (Sketch of experiment described by V. T. Harris)

tors. The two subspecies therefore isolate themselves from each other because of inherited tendencies to choose different parts of a locality for living quarters.

Stanley Wecker raised prairie deer mice in outdoor enclosures that included either natural prairie or forest habitats. When he used the offspring of wild-caught deer mice, their early experience made no difference when he placed them in pens containing a choice of both habitats. The mice chose the prairie habitat when given the opportunity as adults. On the other hand, the offspring of prairie deer mice that had lived in the laboratory for ten to twenty generations did show an effect of early experience, with some tendency to choose the habitat in which they had been brought up. Selection pressure for the proper habitat choice had been relaxed in the generations of laboratory living, and it looks as if there had been a tendency for the mice to "bounce back" in accordance with the principle of genetic homeostasis, meaning that under relaxed selection pressure the mice tended to return to their original behavior before their habitat was differentiated. It is also possible that this is another case of the "Baldwin effect."

The answer to the problem of maintaining species isolation may also be found in the process of socialization. A highly social animal learns early in life that one species is its own kind and others are not, and even in the same locality two species or subspecies could live fairly close together and never mix. This is particularly important among ducks. Many kinds are numerous and widespread, and different species come into repeated contact on the lakes and rivers where they live and breed. Furthermore, the process of socialization and the resulting association of certain individuals tends to isolate subgroups of an originally uniform population. The extent to which this occurs naturally has not yet been thoroughly studied, and it probably varies from species to species. For example, an animal like a buffalo tends to remain in the group of animals with which it grew up. However, there is no very strong limiting mechanism to the size of the herd, which may include several hundred individuals. There is no high degree of inbreeding in such a large group, and we would not expect that rapid evolution would occur. On the other hand, many species of birds regularly divide into small groups during the breeding season. This and the habit of returning to the same geographical locality and breeding territory should produce isolation of small populations. This may account for the great variety of species seen in such passerine birds as the warblers. In the eastern United States there are at least sixteen different species belonging to the genus Dendroica, not counting subspecies, with a great deal of overlap in range.

Another interesting group of birds are three closely related members of the grouse family: the prairie chicken, the sharp-tailed grouse, and the sage grouse. These three species are similar in their general behavior but show an increasing complexity of social organization. The behavior of birds on the mating grounds of the sage grouse is organized so that one dominant cock may do 85 percent of the mating. Since there may be only four or five dominant cocks out of a group of 350, this organization, together with the habit of returning to the same spot each year, results in a considerable degree of inbreeding.

A high degree of social organization of this type should be highly favorable for evolution, and rapid evolution may well have an effect on the development of further social organization. Once a certain degree of organization has been reached, evolutionary change should be rapid, so that there is a good chance of producing an even higher degree of organization. A type of social organization favorable to genetic change should affect the evolution of other characteristics as well, and it is noteworthy that the sage grouse has the most elaborate type of plumage of the three species. It is possible that many of the anatomical traits of birds which seem so bizarre and so difficult to explain on the basis of individual selection can be explained as indirect effects of a social organization favorable for evolution.

Behavior and selection. As we saw earlier in the chapter, the ability to organize behavior on the basis of experience is one which should be favorable to survival, and animals which have it should be selected in favor of those which do not. This situation produces a paradox, for it rapidly gets to the point where an animal is selected for survival on the basis of what it has learned rather than the basic ability to learn. An old and experienced animal has a much better chance to survive than a young one, even if the latter has more inherited ability. The capacity of an animal to adapt thus becomes so remotely related to heredity that it is difficult to see how differential selection could act effectively. Darwin was aware of this difficulty, and in his theory of natural selection he made use of Lamarck's idea of the inheritance of acquired or learned characteristics.

Even great men can make mistakes, and this was in the early days of biological science. Mendel was working on the scientific basis of heredity, but his work was unknown to other scientists, and chromosomes were still to be discovered. Darwin himself was one of the first animal behaviorists, and Pavlov's work on the fundamental facts of learning was still half a century in the future. Darwin quite reasonably made use of the best theory then available to explain the facts.

With our present knowledge we must find another explanation.

Lamarck's hoary theory still raises its head occasionally, even among biologists, but we now know that there is no possible way in which learning can be biologically passed along from one generation to the next, with the possible exception of asexual reproduction in planaria. Learning in animals that reproduce sexually can be transmitted only by means of cultural inheritance, that is, by each generation learning from the preceding one.

This in turn makes it possible for selection to act upon the trait, but through a different level of organization. The social group or population which has a high degree of cultural inheritance has a considerable advantage over a group which does not. This would mean that those populations survive whose members *as a group* have a high degree of the capacity to organize behavior on the basis of experience and thus develop some sort of cultural inheritance.

More than this, we have seen that the welfare of an individual animal is frequently bound up with that of its social group and population. A limited number of heath hens could not withstand the ordinary vicissitudes of nature. Once the population got down to a low level, they were no longer able to reproduce effectively, and the species became extinct. There is every reason for believing that the natural selection of populations may be much more important than the selection of individuals. This gives us a new interpretation of natural animal behavior which seems to fit very well with most of the observed facts.

Fifty years ago it was the fashion to picture the life of animals in nature as a constant battle for survival, with intense individual competition for food, and hungry predators waiting around every corner, ready to snap up the unfit. We have since found that highly competitive situations occur very rarely except in populations which have become disorganized as the result of overcrowding or a disturbed social situation. Even in such cases the deaths seem to be very unselective, with healthy and unhealthy animals all dying together. As for the predators, they often lead lives that are the opposite of the bloody, slavering animals of fiction. We can watch the behavior of coyotes for days without ever seeing them kill a single living thing, and when their stomachs are examined it is evident that a coyote has to eat almost anything that it can get hold of: carrion from animals which have died of disease, garbage, old scraps of leather, and even berries. They do

Fig. 37. Coyote hunting a mouse. These carnivorous mammals spend many hours hunting small animals and are frequently unsuccessful. As a result they often eat carrion and plant food as well as living prey. (After drawing by O. J. Murie, in "The Ecology of the Coyote" by Adolph Murie. U.S. Govt. Printing Office)

occasionally capture small rodents and sometimes are able to find an unprotected newborn fawn. One of the few cases in which coyotes have actually killed an adult deer is so remarkable that it has been written up as a special scientific paper.

The usual coyote hunt may go like this, based on an actual observation: The coyote pair ranges all day several miles from its den. Toward evening, as one of them comes home across the flats, it sees a ground squirrel standing at the entrance of its burrow. The coyote stops, drops to the ground, and begins a long and elaborate stalk, crawling painfully across the open ground. The ground squirrel has meanwhile given its alarm call and is watching the coyote intently. After half an hour or so, the coyote gets to within about thirty feet. But now it attracts the attention of a group of antelope, which bunch together and sweep around behind it, running and circling within thirty or forty feet. The coyote's attention is distracted; it moves its head suddenly and then makes a dash at the ground squirrel, which ducks down its hole in plenty of time. The coyote begins to dig at the hole, and after another half-hour of fruitless effort it gets down a couple of feet. At this point it apparently gives up, stops digging, and lies down on the burrow with its head resting on the pile of dirt. A few minutes later it gets up and trots away.

The animals which are the natural prey of the coyote are so well protected by their behavior and their social groups that it is only rarely that one is caught, usually because it is sick or is a young animal which has accidentally escaped protection. There is undoubtedly some selection against the unwary individual whose behavior for some reason is not coordinated with his protective group. However, this sort of selection would simply lead to the stabilization of the social organization and population, and any real change would depend on differential selection between populations as wholes.

The interaction of selection pressures. In chapter 6 we pointed out six kinds of organized systems that modify and guide the course of development. Each of these forms a source of selection pressure which is more or less independent of the others. Selection pressure from the gene pool eliminates those genes whose action is incompatible with that of the others. Similarly, any gene or combination of genes that is incompatible with the cytoplasm of the egg does not survive. The prenatal environment, in the case of animals that bear their young alive, also selects for gene combinations that survive best in this environment and indeed, the highest death rate of any period of life, something like 40 percent, occurs in this period in the life of mammals. In postnatal life the social, biotic, and physical environments each exert selection pressures.

On what are these pressures exerted? It is obvious that the first three concern individual survival. However, once animals develop sufficiently to become capable of adaptive behavior they become dependent upon each other for survival, even if their contacts are limited to sexual reproduction or an occasional aggregation under unfavorable environmental conditions. Among the highly social animals this interdependence becomes absolutely necessary for survival. For example, a solitary honey-bee cannot live for more than a few days. Furthermore, most species of animals do not exist in large homogeneous populations that include the whole species. Rather, they are organized into subpopulations, sometimes on the basis of geographical barriers, and often within these barriers on the basis of social groups. These groups may be as small as a mated pair and their offspring, as in certain of the birds of prey, or as large as a troop of monkeys or even as a much larger herd of elk.

From the viewpoint of genetics, natural selection determines only incidentally the survival of an individual or a group. What is important is the survival of a gene or a particular combination of genes. The fact that an individual survives has no importance if it is unable to mate and insure the survival of its genes. Consequently the survival of a group is as important as that of the individual. As Wynne-Edwards points out, there can be differential survival of subpopulations and social groups as well as of individuals within those groups.

Because there are many different kinds of selection pressures, and because group survival and individual survival are not necessarily equivalent, it is possible for different selection pressures to operate in opposite ways. Theoretically, there are four different combinations of positive and negative effects of selection exerted against groups and individuals. In the first combination, group survival and individual survival are both promoted by a particular kind of behavior. For example, the allelomimetic behavior in a wolf pack in the forms of cooperative hunting and common defense of dens against predators promotes both group and individual survival. We would therefore predict that selection pressure would strongly favor the continuance of this behavior. Second, selection pressures can operate negatively for an individual and positively for a group. This has been compared to altruistic behavior in human beings, and an example is the "broken wing" behavior pattern of the parent birds in certain ground-nesting species. When a potential predator comes near, the adult killdeer flutters along the ground as if a wing were broken, first distracting the predator's attention away from the nest or young birds, and then flying away unharmed. Contrarily, selection may act positively for individual survival and negatively for a group. An example is the competition over food seen in packs of wolves. The individual who gets most of the food will survive better, but to the detriment of his pack mates. In both these cases we would predict that opposite selection pressures would result in some sort of equilibrium which would promote maximum survival of both individual and group. The "broken wing" behavior of the killdeer must not be too realistic or both the parent and young will die. The wolf cannot be too greedy or his pack mates will not survive to help him in the hunt. Indeed, wolves normally share food as well as competing for it. An individual that has made a kill will howl and attract his pack mates to the scene.

Finally, there is the theoretical case where a particular pattern of behavior has negative survival value for both group and individual. We would predict that such behavior would be strongly selected against. An example is destructively violent agonistic behavior resulting in death or serious injury to the combatants. Studies of agonistic behavior under normal ecological conditions and social organization show that there is a general evolutionary tendency to reduce social fighting to relatively nonharmful forms. Furthermore, it is extremely rare to have any sort of fighting between opposite sexes that would interfere with mating.

What kinds of selection are most important for behavioral evolution? There is no general answer, as this depends upon the particular ecosystem and type of social organization of the species. For example, in Australia the marsupial mammals never evolved any highly successful predators. Consequently the mallee fowl has largely been affected by selection pressures from the physical environment, and much of its behavior is devoted to coping with and controlling such environmental changes. Its existence has now been threatened by the introduction of foxes, to which both the birds and their nests are highly vulnerable. For this and other species that are relatively solitary, selection pressures from the biotic and physical environment are most important. As a species becomes more social, it offers more and more protection from these sorts of selection pressures, with the result that social selection pressure becomes increasingly important. For a highly social animal, the most important part of survival is survival in and adjustment to a social group. Under these conditions social selection may guide the evolution of behavior into channels that would be destructive to a solitary individual. As we noted above, the territorial songs and displays and other social signals evolved by the passerine birds would be completely nonadaptive to a nonsocial species, as they would only serve to attract the attention of predators. A similar example is the evolution of agonistic behavior in the deer family. The pushing contests between males in the breeding season determine which males will do the greatest amount of mating. Since larger and heavier antlers are an advantage, there has been an evolutionary tendency for these to become bigger and bigger, although they are useless outside the breeding season, are discarded shortly after it, and are a handicap as they grow in the velvet. Thus an organ which has great selective advantage in a social situation is produced by the individual at the price of a considerable outlay of energy and a sacrifice of survival capacities under nonsocial conditions.

The evolution of behavior can therefore be understood only

with reference to the particular kinds of selection pressures that operate upon a given species and the particular gene pool that is available to it as the result of previous evolution. Among these pressures we must give more and more attention to that of social selection, using the term in a broader sense than Darwin's concept of sexual selection. While many species, like the deer, do compete for mates, the social environment includes much more than agonistic and sexual behavior.

There are no universal rules for behavioral evolution. Each species evolves independently along a unique path and does not represent, as was once thought, a stage in the evolution of human behavior. Comparisons with other species enable us to measure the evolution of human behavior more exactly, but they do not provide precise analogues.

Human evolution. The body structures of both modern men and the newly discovered fossils from Africa indicate that early man was a plains-living primate who used caves for shelter and roamed over the surrounding countryside gathering food. If these men were divided into relatively small social groups, each inhabiting a particular area and having little interchange between the populations, a situation favorable to rapid evolution would be present. Once verbal communication had appeared in one of these groups, the selective advantages over any other group would be so great as to insure its survival as the others disappeared. A language also makes possible more definite and separate social units, since it excludes outsiders who do not speak it. Small bands of human beings, partially isolated by language, would fit the conditions of rapid evolutionary change very well indeed.

Social organization leaves no fossils, only artifacts, and we can only check these assumptions with observations on primitive tribes which have survived into historical times in the deserts of Australia and Africa and in the mountains of New Guinea. Their dialects and customs vary from one area to the next, so that few people change from one tribe to another, and the tribes have populations ranging from a few hundred up to 2,000 members. This is a most favorable size for rapid genetic change, since an extremely small population is as unfavorable for evolution as a very large one. Rapid evolution of the biological ability for learning and speaking a verbal language would be expected under these circumstances, as well as changes in other traits.

Once a high degree of verbal ability has been established, it becomes almost entirely independent of biological control, and human societies undergo a kind of cultural evolution which has little relation to biological heredity. In fact, at the present time cultural evolution has proceeded in a direction which is unfavorable to biological change. The large and shifting populations in modern civilized societies very nearly meet the basic assumptions for Hardy's law: an infinitely large population with random mating. At the same time, increasingly efficient social organization protects us from certain kinds of selection pressures, with a resulting increase in individual variation. This plus assortive mating within societies makes for an evolutionary trend in which there is little change in mankind's general average, lessening differences between subpopulations, and enormously increasing individual variation.

THE SCIENCE OF ANIMAL BEHAVIOR

The evolution of any species, and particularly of a highly social species, cannot be understood without studying its behavior and social organization. Evolution is one of the fundamental theories of biology. Its basis is adaptation, and one of the important kinds of adaptation is behavior. The earlier chapters of this book outline the various research problems of behavioral adaptation. The ramifications of this research extend into almost every corner of biology and into some related sciences like psychology and sociology. Even the physical sciences are brought in to explain the physiology of behavior, and mathematics is used in the study of populations.

We have derived certain basic principles from the facts of individual animal behavior and have developed from them broader theories and hypotheses which explain the organization and behavior of social groups and populations. The results are impressive, but the need for further work is even more impressive. Very few wild species have been thoroughly studied from the behavioral point of view, and we still cannot say that we thoroughly understand every major system of behavior in even a single species. Both the fundamental process of learning and the physiology of behavior have been studied in connection with only a few patterns of behavior and on a small number of species. The way in which heredity produces individual differences within a species is still not adequately understood.

We need to extend our observations and experiments to a wide variety of species, after which it may be possible to state general laws and theories with a great deal more certainty and to use them as a real foundation for human knowledge. Progress is now being made in many directions: the genetics of behavior, its adaptive organization in the central nervous system, and the organization of groups through communication. Some of the most interesting and hopeful advances are coming from the study of animal behavior in relation to social organization and populations. At the present time the systematic study of the behavior of any species brings out new facts and sometimes startling new ideas. There are periods in the history of any science when conditions are right for rapid advances, and, although we here come to the end of this book, we still stand at the beginning of the science of animal behavior.

Bibliography

The "General Reading and References" lists for each chapter include books and articles that are either unusually well written or can be used as references for looking up detailed information. The "Additional References" lists include technical source material that has been specifically referred to in the text. Where a reference applies to more than one chapter, it is listed only under the chapter where it is first used. The references chiefly include works available in English or English translation; the advanced student can use these as a guide to articles and books published in other languages.

Chapter 1

GENERAL READING AND REFERENCES

Jennings, H. S. 1906. *Behavior of the lower organisms.* New York: Columbia Univ. Press.

A basic book on the general facts which support the stimulus-response theory. Written in 1906, it is still one of the best references on the behavior of protozoans and coelenterates.

Klopfer, P. H., and Hailman, J. P. 1967. *An introduction to animal behavior: Ethology's first century.* Englewood Cliffs, N.J.: Prentice-Hall.

General introduction from a historical viewpoint; best as a guide to further reading.

SCIENTIFIC JOURNALS

The latest reports of new data on animal behavior are to be found in various scientific journals. The short articles in *Science* and the English magazine *Nature* provide the most rapid means of publication and are most up-to-date. *Science* also frequently carries lead review articles which are usually written in comprehensive style. *Scientific American* also frequently includes general

301

articles written in nontechnical language by well-known authorities, but these are not documented and hence cannot be examined for critical evidence.

The most widely read journal devoted exclusively to the subject is the British-American journal *Animal Behaviour.* This journal also publishes longer monographs occasionally. Another journal exclusively devoted to the subject is *Behaviour,* which has an international board of editors and publishes most of its articles in English. This journal usually includes longer and more detailed descriptive articles. Corresponding to *Animal Behaviour* in Germany is the *Zeitschrift für Tierpsychologie,* where the papers are published with English summaries.

In addition, many papers are published according to taxonomic classification, so that papers on mammalian behavior appear in *Mammalogy* and those on bird behavior appear in the *Auk* and other ornithological journals. Papers on insect behavior are likely to appear in various journals devoted to entomology. Other papers on animal behavior may be found in widely scattered biological journals. Comprehensive indexes can be found in *Biological Abstracts* and *Psychological Abstracts.* Another useful abstract journal is *Wildlife Review.*

In addition, there are certain new specialized journals on animal behavior: *Developmental Psychobiology, Behavior Genetics,* and *The International Journal of Psychobiology.*

American psychologists have no journals specializing in animal behavior, but experimental papers frequently appear in *The Journal of Genetic Psychology* and *The Journal of Comparative and Physiological Psychology.*

In addition, many of the outstanding articles on animal behavior have been reprinted in book form as collections of selected readings.

Chapter 2

RECOMMENDED GENERAL READING

Darling, F. F. 1937. *A herd of red deer.* Oxford: Clarendon Press. This interesting "Study in Animal Behavior" contains a vivid description of the technique of studying a large mammal under field conditions.

Ficken, R. W. 1963. Courtship and agonistic behavior of the common grackle, *Quiscalus quiscula. Auk* 80:52–72.
Modern study of the behavior patterns of a particular bird species.

Goodall, J. 1965. Chimpanzees of the Gombe Stream Reserve. In *Primate behavior: Field studies of monkeys and apes,* ed. I. Devore, chap. 12. New York: Holt, Rinehart & Winston.
Demonstrates the value of the technique of habituating animals to a human observer.

Hafez, E. S. E., ed. 1969. *The behaviour of domestic animals.* 2d ed. London: Ballière, Tindall & Cassell.
A general reference book on animal behavior as illustrated by domestic animals. Ingestive and sexual behavior are emphasized because of their economic importance.

Jay, P. 1965. The common langur of North India. In *Primate behavior: Field studies of monkeys and apes,* ed. I. Devore, chap. 7. New York: Holt, Rinehart & Winston.
A field study under natural conditions, particularly good for its observations of development of behavior.

Nice, M. M. 1937; 1943. Studies in the life history of the song-sparrow. *Transactions of the Linnaean Society of New York* 4:1–247; 6:1–328.
These well-written volumes should be read by anyone seriously interested in bird behavior. In addition to a detailed description of the behavior of the song sparrow, the author compares it with many other species and gives references to all the important advances in scientific bird study, such as socialization, song, and territoriality, referred to in later chapters of this book. The song sparrow shows essentially the same general types of adaptation as does the red-winged blackbird used as an example in this chapter.

Schaller, G. B. 1963. *The mountain gorilla.* Chicago: Univ. of Chicago Press.
A major study of behavior under natural conditions.

Scott, J. P., ed. 1950. Methodology and techniques for the study of animal societies. *Annals of the New York Academy of Sciences,* 51:1001–1122.
Contains papers by several authors who present their ideas of

the best and most approved methods for studying animal behavior. Emlen gives general methods for the study of birds, and Schneirla gives the theoretical reasons for the methods used on all animals. His section on "Observational control through record taking" is particularly good. J. W. Scott illustrates the comparative method with a study of three types of grouse, and J. P. Scott has a similar comparative paper on dogs and wolves. The sections on technical apparatus are now out of date.

Stokes, A. W., ed. 1968. *Animal behavior in laboratory and field.* San Francisco: W. H. Freeman.
A variety of laboratory and field exercises contributed by members of the Animal Behavior Society and intended for use in formal courses.

Tinbergen, N. 1965. Behavior and natural selection. In *Ideas in modern biology,* ed. J. A. Moore. Garden City, N.Y.: Natural History Press.
Argues that adaptive value of behavior can be tested experimentally.

ADDITIONAL REFERENCES

Allen, A. A. 1911–13. The red-winged blackbird: A study in the ecology of a cat-tail marsh. *Abstract Proceedings of the Linnaean Society of New York,* pp. 43–128.

Aschoff, J., ed. 1964. Circadian clocks: Proceedings of the Feldafing Summer School. Amsterdam: North Holland Publishing Co.

Emlen, J. T. 1952. Social behavior in nesting cliff swallows. *Condor* 54:177–99.

Nero, R. W. 1956. A behavior study of the red-winged blackbird. *Wilson Bulletin* 68:5–37, 129–50.

Schneider, D. E. 1968. Post-breeding season flocking and roosting behavior of the red-winged blackbird (*Agelaius phoenecius*). Master's thesis, Bowling Green University.

Scott, J. P. 1945. Social behavior, organization and leadership in a small flock of domestic sheep. *Comparative Psychology Monographs* 18(4):1–29.

Shaw, Evelyn. 1969. Schooling in fishes: Critique and review. In *The development and evolution of behavior,* ed. L. R. Aronson, E. Tobach, J. S. Rosenblatt, and D. S. Lehrman. San Francisco: W. H. Freeman.

Chapter 3
RECOMMENDED GENERAL READING

Collias, N. E., organizer. 1964. Symposium on the Evolution of external construction by animals. *American Zoologist* 4:175–243.
Papers on nest-building in birds, weaving of spider webs, and other examples of complex motor behavior.

Ingle, D., ed. 1968. *The central nervous system and fish behavior.* Chicago: Univ. of Chicago Press.
Contains an excellent section on fish behavior.

Marler, P. R., and Hamilton, W. J. III. 1966. *Mechanisms of animal behavior.* New York: Wiley.
General survey of literature for advanced students, with special emphasis on sensory mechanisms.

Prosser, C. L. 1961. Nervous systems. In *Comparative animal physiology,* ed. C. L. Prosser and F. A. Brown, chap. 23. 2d ed. Philadelphia: W. B. Saunders.
Well-organized reference work for the facts of comparative nervous physiology.

Walls, G. L. 1963. *The vertebrate eye and its adaptive radiation.* New York: Hafner.
Encyclopedic reference work on the structure and functions of the vertebrate eye.

ADDITIONAL REFERENCES

Black-Cleworth, P. 1970. The role of electrical discharges in the non-reproductive social behavior of *Gymnotus carapo* (Gymnotidae, Pisces). *Animal Behavior Monographs* 3 (1):1–77.

Boycott, B. B., and Young, J. Z. 1950. The comparative study of learning. In *Symposia of the Society for Experimental Biology. IV. Physiological mechanisms in animal behavior,* pp. 432–53. New York: Academic Press.

Butler, R. A., and Harlow, H. F. 1954. Persistence of visual exploration in monkeys. *Journal of Comparative and Physiological Psychology* 47:258–63.

Cohn, P. H., ed. 1967. *Lateral line detectors.* Bloomington, Ind.: Indiana Univ. Press.

Garcia, J.; Buchwald, B. H.; Koelling, R. A.; and Tedrow, L. 1964. Sensitivity of the head to X-ray. *Science* 144:1470–72.

Jones, F. R., and Marshall, N. B. 1953. The structure and functions of the teleostean swimbladder. *Biological Reviews* 28:16–83.

King, J. A. 1956. Social relations of the domestic guinea pig living under semi-natural conditions. *Ecology* 37:221–28.

Chapter 4

RECOMMENDED GENERAL READING

Barnett, S. A. 1963. *The rat: A study in behaviour.* Chicago: Aldine.
Excellent and well-written study of the wild Norway rat, with some emphasis on physiology.

Beach, F. A., ed. 1965. *Sex and behavior.* New York: Wiley.
Contains a variety of technical articles on sex research on non-human animals, with some emphasis on physiological factors.

Cannon, W. B. 1929. *Bodily changes in pain, hunger, fear and rage.* Boston: Branford.
Still an excellent book on the physiological causes of behavior in mammals and particularly in human beings. For more recent research, see Gellhorn.

Collias, N. E. 1944. Aggressive behavior among vertebrate animals. *Physiological Zoology* 17:83–123.
A general review, including a section on physiological mechanisms affecting fighting behavior.

Gellhorn, E. 1953. *Physiological foundations of neurology and psychiatry.* Minneapolis: Univ. of Minnesota Press.
Chap. 14, "The physiological basis of emotion," contains an authoritative and comprehensive summary of experimental work on the nervous systems of brains of mammals.

Manning, A. 1967. *An introduction to animal behavior.* Reading, Mass.: Addison-Wesley.
A well-written introduction to the physiology of animal behavior.

Scott, J. P. 1958. *Aggression.* Chicago: Univ. of Chicago Press.
A general summary of factors inducing agonistic behavior, with emphasis on physiology, genetics, and experience.

———. 1967. Agonistic behavior in mice and rats: A review. *American Zoologist* 7:373–81.
Review of the various kinds of external and internal stimulation that affect the agonistic behavior of these two species.

Southwick, C. H., ed. 1970. *Animal aggression: Selected readings.* New York: Van Nostrand-Reinhold.
Well-balanced group of papers illustrating varying views of authorities on the causes of social fighting.

Stellar, E. 1954. The physiology of motivation. *Psychological Review* 61:5–22.
Particularly good on the function of the hypothalamus.

Thompson, R. F. 1967. *Foundations of physiological psychology.* New York: Harper & Row.
Good modern reference book on neurophysiology and behavior.

Tobach, E., ed. 1969. Experimental approaches to the study of emotional behavior. *Annals of the New York Academy of Sciences* 159:621–1121.
Summaries of modern research on emotions in nonhuman animals.

Young, W. C., ed. 1961. *Sex and internal secretions.* 3d ed. Baltimore, Md.: Williams and Wilkins.
A standard reference work on the physiology of sex.

ADDITIONAL REFERENCES

Almquist, J. O., and Hale, E. B. 1956. An approach to the measurement of sexual behavior and semen production of dairy bulls. *Proceedings 3rd International Congress on Animal Reproduction* (Cambridge, England), Plenary Papers, pp. 50–59.

Bard, P. 1950. Central nervous mechanisms for the expression of anger in animals. In *Feelings and emotions,* ed. M. L. Reymert. New York: McGraw-Hill.

Bard, P., and Mountcastle, V. B. 1948. Some forebrain mechanisms involved in expression of rage with special reference to suppression of angry behavior. *Proceedings of the Association for Research in Nervous and Mental Disease* 27:362–404.

Beeman, E. A. 1947. The effect of male hormone on aggressive behavior in mice. *Physiological Zoology* 20:373–405.

Bronson, F., and Desjardins, C. 1971. Steroid hormones and aggressive behavior in animals. In *The physiology of aggression and defeat*, ed. B. E. Eleftheriou and J. P. Scott. New York, Plenum Press.

Bronson, F. H., and Eleftheriou, B. E. 1964. Chronic physiological effects of fighting in mice. *General and Comparative Endocrinology* 4:9–14.

Chambers, R. M. 1956. Effects of intravenous glucose injections on learning, general activity, and hunger drive. *Journal of Comparative and Physiological Psychology* 49:558–64.

Eleftheriou, B. E., and Scott, J. P., eds. 1971. *The physiology of aggression and defeat*. New York, Plenum Press.

Fisher, A. E. 1964. Chemical stimulation of the brain. *Scientific American* 210:60–68.

Guhl, A. M. 1961. Gonadal hormones and social behavior in infrahuman vertebrates. In *Sex and internal secretions*, ed. W. C. Young. Baltimore: Williams and Wilkins Co.

Hess, W. R. 1954. *Diencephalon: Autonomic and extrapyramidal functions*. New York: Grune & Stratton.

Kaada, B. 1967. Brain mechanisms related to aggressive behavior. In *Aggression and defense: Neural mechanisms and social patterns*, ed. C. D. Clemente and D. B. Lindsley. Los Angeles: Univ. of California Press.

Kendeigh, S. C. 1952. *Parental care and its evolution in birds*. Illinois Biological Monographs 22:1–343. Urbana: Univ. of Illinois Press.

Lehrman, D. S. 1965. Interaction between internal and external environments in the regulation of the reproductive cycle of the ring dove. In *Sex and behavior*, ed. F. Beach. New York: Wiley.

Levine, S., and Mullins, R. F., Jr. 1966. Hormonal influences on brain organization in infant rats. *Science* 152:1585–92.

Ranson, S. W. 1934. The hypothalamus: Its significance for visceral innervation and emotional expression. *Transactions of the College of Physicians of Philadelphia* 2:222–42.

Richter, C. P. 1947. Biology of drives. *Journal of Comparative and Physiological Psychology* 40:129–34.

Roberts, W. W.; Steinberg, M. L.; and Means, L. 1967. Hypothalamic mechanisms for sexual, aggressive and other motivational behaviors in the opossum, *Didelphis Virginiana*. *Journal of Comparative and Physiological Psychology* 64:1–15.

Scott, E. M., and Quint, E. 1946. Self-selection of diet. IV. Appetite for protein. *Journal of Nutrition* 32:293–302.

Schachter, S. 1964. The interaction of cognitive and physiological determinants of emotional state. In *Psychobiological Approaches to Social Behavior*, ed. P. H. Leiderman and D. Shapiro. Stanford: Stanford Univ. Press.

Selye, H. 1950. *Stress*. Montreal: Acta, Inc.

Terkel, J., and Rosenblatt, J. S. 1968. Maternal behavior induced by maternal blood plasma injected into virgin rats. *Journal of Comparative and Physiological Psychology* 65:479–82.

Tollman, J., and King, J. 1956. The effects of testosterone propionate on aggression in male and female C57BL/10 mice. *British Journal of Animal Behaviour* 4(4):147–49.

Chapter 5

RECOMMENDED GENERAL READING

Corning, W. C., and Ratner, S. C., eds. 1967. *Chemistry of learning: Invertebrate research*. New York: Plenum Press.
Chiefly technical papers on learning in planaria; also chapters on learning in other invertebrates.

Hilgard, E. R., and Bower, G. H. 1966. *Theories of learning*. New York: Appleton-Century-Crofts.
A standard reference work on the psychological theories of learning.

Maier, W. R. F., and Schneirla, T. C. 1935. *Principles of animal psychology*. New York: McGraw-Hill.
Presents in detail the older evidence of learning abilities found in animals from protozoans to man.

Munn, N. L. 1950. *Handbook of psychological research on the rat.* New York: Houghton Mifflin Co.
Includes an excellent summary of the various theories of learning which have been developed so largely from the behavior of the rat. Also a good reference book for looking up experimental techniques and apparatus which have been used on this animal.

Pavlov, I. P. 1927. *Conditioned reflexes.* London: Oxford University Press; Humphrey Milford.
Not easy reading, but gives many examples of the original research on conditioned reflexes. Pavlov's neurological explanations of learning now have little importance, but his well-established facts form the foundation for much of the theorizing done by later workers.

Skinner, B. F. 1938. *The behavior of organisms.* New York: Appleton-Century-Crofts.
Skinner's original work on operant conditioning.

Thorpe, W. H. 1963. *Learning and instinct in animals.* 2d ed. London: Methuen.
Best reference on learning in animals other than mammals.

———. 1965. The ontogeny of behavior. In *Ideas in modern biology,* ed. J. A. Moore. Garden City, N.Y.: Natural History Press.
Good summary on learning, especially habituation.

Warden, C. J.; Jenkins, T. W.; and Warner, L. H. 1936. *Comparative psychology.* 3 vols. New York: Ronald Press.
A systematic reference book for descriptions of experimental work on the sensory, motor, and learning capacities of any species, with an extensive bibliography.

ADDITIONAL REFERENCES

Ginsburg, B., and Allee, W. C. 1942. Some effects of conditioning on social dominance and subordination in inbred strains of mice. *Physiological Zoology* 15:485–506.

Hovey, H. B. 1929. Associative hysteresis in marine flatworms. *Physiological Zoology* 2:322–33.

Liddell, H. S. 1944. Animal behavior studies bearing on the problem of pain. *Psychosomatic Medicine* 6:261–65.

McConnell, J. V.; Jacobson, A. L.; and Kimble, D. P. 1959. The effects of regeneration upon retention of a conditioned response in the planaria. *Journal of Comparative and Physiological Psychology* 52(1):1–5.

Nissen, H. W. 1951. Phylogenetic comparison. In *Handbook of experimental psychology*, ed. S. S. Stevens, chap. 11. New York: John Wiley & Sons.

Rosenblatt, J. S., and Aronson, L. R. 1958. The influence of experience on the behavioural effects of androgen in prepuberally castrated male cats. *Animal Behaviour* 6:171–82.

Scott, J. P., and Marston, M. V. 1953. Nonadaptive behavior resulting from a series of defeats in fighting mice. *Journal of Abnormal and Social Psychology* 48:417–28.

Solomon, R. L.; Kamin, L. J.; and Wynne, L. C. 1953. Traumatic avoidance learning: The outcomes of several extinction procedures with dogs. *Journal of Abnormal and Social Psychology* 48:291–302.

Chapter 6

RECOMMENDED GENERAL READING

Banks, E. M., ed. 1967. Ecology and behavior of the wolf. *American Zoologist* 7:220–381.
A collection of papers by authorities on wolf behavior summarizing recent information on this species.

Bastock, M. 1967. *Courtship: An ethological study.* Chicago: Aldine.
Evolution and genetics of preliminary sexual behavior in animals other than mammals.

Bliss, E. L., ed. 1962. *Roots of behavior: Genetics, instinct and socialization in animal behavior.* New York: Harper.
Chapters on basic general problems in animal behavior written by authorities in the field.

Dobzhansky, T. 1964. *Heredity and the nature of man.* New York: Harcourt, Brace & World.
Conclusions concerning the biological nature of man derived from modern genetics.

Fuller, J. L., and Thompson, W. R. 1960. *Behavior genetics.* New York: Wiley.
A standard textbook in the field, relating results with human and nonhuman animals.

Scott, J. P., and Fuller, J. L. 1965. *Genetics and the social behavior of the dog.* Chicago: Univ. of Chicago Press.
Summarizes the results of a thirteen-year long experiment with five dog breeds and hybrids between two of them.

ADDITIONAL REFERENCES

Caspari, E.; Fuller, J. L.; Rothenbuhler, W. C.; Bruell, J. H.; Hirsch, J.; Ehrman, L.; Dilger, W. S.; and Scott, J. P. 1964. Refresher course in behavior genetics. *American Zoologist* 4: 97–173.

Collins, R. L., and Fuller, J. L. 1969. Genetic and environmental factors in audiogenic seizure susceptibility. Paper delivered at AAAS, Boston.

Craig, J. V.; Ortman, L. L.; and Guhl, A. M. 1965. Genetic selection for social dominance ability in chickens. *Animal Behaviour* 13:114–31.

Crew, F. A. E. 1927. Abnormal sexuality in animals. III. Sex reversal. *Quarterly Review of Biology* 2:427–41.

———. 1946. *Sex determination.* London: Methuen.

David, P. R., and Snyder, L. H. 1951. Genetic variability and human behavior. In *Social Psychology at the Crossroads,* ed. J. H. Rohrer and M. Sherif, chap. 3. New York: Harper & Bros.

Dice, L. R. 1935. Inheritance of waltzing and of epilepsy in mice of the genus *Peromyscus. Journal of Mammalogy* 16:25–35.

Dilger, W. 1962. The behavior of lovebirds. *Scientific American* 206:88–98.

Fredericson, E. 1952. Reciprocal fostering of two inbred mouse strains and its effect on the modification of inherited aggressive behavior. *American Psychologist* 7:241–42.

Fuller, J. L., and Scott, J. P. 1954. Heredity and learning ability in infrahuman mammals. *Eugenics Quarterly* 1:28–43.

Ginsburg, B. E. 1967. Genetic parameters in behavioral research. In *Behavior-genetic analysis,* ed. J. Hirsch. New York: McGraw-Hill.

Hall, C. S. 1941. Temperament: A survey of animal studies. *Psychological Bulletin* 38:909–43.

———. 1947. Genetic differences in fatal audiogenic seizures between two inbred strains of house mice. *Journal of Heredity*, 38:2–6.

Hirsch, J. 1959. Studies in experimental behavior genetics: II. Individual differences in geotaxis as a function of chromosome variations in synthesized *Drosophila* populations. *Journal of Comparative and Physiological Psychology* 52:304–8.

Honess, R. F., and Frost, N. M. 1942. A Wyoming bighorn sheep study. Wyoming Fish and Game Department Bulletin no. 1.

Hudgens, G. A.; Denenberg, V. H.; and Zarrow, M. X. 1968. Mice reared with rats: Effects of pre-weaning and post-weaning social interactions upon adult behavior. *Behaviour* 30:259–74.

Potter, J. H. 1949. Dominance relations between different breeds of domestic hens. *Physiological Zoology* 22:261–80.

Scott, J. P. 1943. Effects of single genes on the behavior of *Drosophila*. *American Naturalist* 77:184–90.

Speicher, B. M. Males in *Nemeritis*. Private communication.

Spuhler, J. N., ed. 1967. *Genetic diversity and human behavior*. Chicago: Aldine.

Vandenberg, S. G., ed. 1965. *Methods and goals in human behavior genetics*. New York: Academic Press.

Whiting, P. W. 1932. Reproductive reactions of sex mosaics of a parasitic wasp, *Habrobracon juglandis*. *Journal of Comparative Psychology* 14:345–63.

Yerkes, R. M. 1907. *The dancing mouse*. New York: Macmillan Co.

Chapter 7

RECOMMENDED GENERAL READING

(See also general references under chap. 6.)

Eibl-Eibesfeldt, I. 1970. *Ethology: The biology of behavior*. Trans. E. Klinghammer. New York: Holt, Rinehart and Winston.
General review of animal behavior emphasizing instinct theory.

Hebb, D. O. 1949. *The organization of behavior*. New York: John Wiley & Sons.
Contains many stimulating ideas concerning the physiological mechanisms behind the organization of behavior.

Hinde, R. A. 1970. *Animal behavior: A synthesis of ethology and comparative psychology*, 2d ed. New York: McGraw-Hill.
An attempt to reconcile divergent viewpoints of psychologists and biologists, with emphasis on the problem of motivation.

Katz, D. 1937; 1953. *Animals and men*. New York: Longmans, Green; London: Penguin Books.
Describes many interesting experiments done by European animal behaviorists and not usually available in English. Included among others are Clever Hans and the Elberfeld horses.

Koehler, W. 1927. *The mentality of apes*. 2d ed. New York: Harcourt-Brace.
Koehler's account of his classical experiments on the organization of behavior in chimpanzees.

Krushinski, L. V. 1962. *Animal behavior: Its normal and abnormal development*. New York: Consultants Bureau.
Introduction to Russian experiments on animal behavior, with emphasis on the dog and employing Pavlovian terminology. Some experiments with problem-solving (here called extrapolation reflexes) employ unusual techniques.

Kuo, Z. 1967. *The dynamics of behavior development*. New York: Random House.
A summary of the author's work containing a critique of the concept of instinct.

Lack, D. 1946; 1953. *The life of the robin*. London: H. F. & G. Witherby; rev. ed. London: Penguin Books.
The ecology and behavior of the English robin. Chap. 12 describes Lack's interesting experiments with stuffed models.

Newton, G., and Levine, S., eds. 1968. *Early experience and behavior*. Springfield, Ill.: Charles C. Thomas.
Collection of technical review papers on a variety of early experience phenomena.

Schiller, C. H., ed. 1957. *Instinctive behavior: The development of a modern concept*. New York: International Universities Press.

Articles on the theory of instinct by Lorenz, Tinbergen, von Uexküll, and others.

Scott, J. P. 1968. *Early experience and the organization of behavior*. Belmont, Calif.: Wadsworth Publishing Co.
Presents a more extended review of the effects of modifying the various preorganized systems that affect the development of behavior, with some emphasis on the critical period phenomenon.

Tembrock, G. 1964. Verhaltensforschung: Eine Einführung in die Tier-Ethologie. Jena: G. Fischer.
Good general reference in German. Covers English as well as German literature on animal behavior.

Tinbergen, N. 1951. *The study of instinct*. Oxford: Clarendon Press.
An authoritative book on the comparative study of behavior patterns and their evolution.

ADDITIONAL REFERENCES

Fraenkel, G. S., and Gunn, D. L. 1964. *The orientation of animals*. Oxford: Clarendon Press.

Fuller, J. L., and Clark, L. D. 1966. Genetic and treatment factors modifying the post-isolation syndrome in dogs. *Journal of Comparative and Physiological Psychology* 61:251–57.

Fuller, J. L., and Clark, L. D. 1966. Effects of rearing with specific stimuli upon post-isolation behavior in dogs. *Journal of Comparative and Physiological Psychology* 61:258–63.

Hall, K. R. L. 1963. Tool using performances as indicators of behavioral adaptability. *Current Anthropology* 4(5):479–94.

Hall, K. R. L., and Schaller, G. B. 1964. Tool using behavior of the California sea otter. *Journal of Mammalogy* 45:287–98.

Harlow, H. F. 1958. The nature of love. *American Psychologist* 13:673–85.

Hebb, D. O. 1947. The effects of early experience on problem-solving at maturity. *American Psychologist* 2:306–7.

Kortlandt, A. 1967. Experimentation with chimpanzees in the wild. In *Progress in Primatology*, ed. D. Starck, R. Schneider, and H. S. Kuhn. Stuttgart: Gustav Fischer.

Loeb, J. 1918. *Forced movements, tropisms, and animal conduct*. New York: Lippincott.

Scott, J. P. 1962. Critical periods in development. *Science* 138: 949–58.

———. 1967. The development of social motivation. In *Nebraska symposium on motivation,* ed. D. Levine. Lincoln, Neb.: Univ. of Nebraska Press.

Thompson, W. R., and Heron, W. 1954. The effects of restricting early experience on the problem-solving capacities of dogs. *Canadian Journal of Psychology* 8:17–31.

Tinbergen, N. 1953. *The herring gull's world.* London: Collins.

Chapter 8

RECOMMENDED GENERAL READING

Altmann, S., ed. 1967. *Social communication among primates.* Chicago: Univ. of Chicago Press.
Field and laboratory studies on social behavior of a large variety of primates.

Carpenter, C. R. 1964. *Naturalistic behavior of non-human primates.* University Park: Pennsylvania State Univ. Press.
Reprinted edition of the author's classical monographs on howling monkeys and gibbons, as well as papers on other primates.

Devore, I., ed. 1965. *Primate behavior.* New York: Holt, Rinehart & Winston.
Field studies on a wide variety of monkeys and apes.

Etkin, W., ed. 1964. *Social behavior and organization among vertebrates.* Chicago: Univ. of Chicago Press.
Well-organized general book, with special chapters by a number of authorities.

Grasse, P. P., ed. 1952. *Structure et physiologie des sociétés animales.* Paris: Colloques internationaux du centre national de la recherche scientifique.
Contains several interesting articles on animal societies by such authorities as Allee, Carpenter, Darling, Emerson, and Schneirla, besides many others in French and German by European authorities.

Guhl, A. M. 1953. *Social behavior of the domestic fowl.* Kansas State College Agricultural Experiment Station Technical Bulletin no. 73.

An excellent summary of research work on the peck order of hens.

Jay, P., ed. 1968. *Primates: Studies in adaptation and variability.* New York: Holt, Rinehart & Winston.
Particularly interesting from the viewpoint of variation in social organization both within and between different species.

Jolly, A. 1966. Lemur social behavior and primate intelligence. *Science* 153:501–6.
An account of the social behavior of these primitive primates.

Kummer, H. 1968. *Social organization of hamadryas baboons.* Chicago: Univ. of Chicago Press.
Modern field study of the species studied by Zuckerman.

Lorenz, K. 1969. *Studies in animal and human behaviour,* vol. 1. Boston: Harvard Univ. Press.
English translation of some of Lorenz's major papers on bird behavior.

McBride, G. 1964. A general theory of social organization and behaviour. *University of Queensland Papers, Faculty of Veterinary Science* 1(2):75–110.
General paper on theory of social organization.

Murchison, C. A. 1935. *Handbook of social psychology.* Worcester, Mass.: Clark Univ. Press.
This book, now out of print, contains chapters by many noted authorities, including one by Schjelderup-Ebbe on the peck order in chickens.

Portmann, A. 1961. *Animals as social beings.* New York: Viking.
A popularly written summary of European research on social behavior of animals.

Rheingold, H. L., ed. 1963. *Maternal behavior in mammals.* New York: Wiley.
Studies on twelve species of mammals belonging to five different orders and emphasizing laboratory and domestic varieties.

Sluckin, W. 1965. *Imprinting and early learning.* Chicago: Aldine.
Good review of research on the formation of primary social attachments.

Southwick, C. H., ed. 1963. *Primate social behavior.* Princeton: Van Nostrand.
Reprints of some of the classical papers on social behavior in primates, particularly monkeys.

Wilson, E. O. 1963. The social biology of ants. *Annual Review of Entomology* 8:345–68.
Review of some of the newer discoveries in ants, with emphasis on pheromones and division of labor.

Zuckerman, S. 1932. *The social life of monkeys and apes.* New York: Harcourt-Brace.
Contains an account of the author's observations on the social life of baboons in the London Zoo.

ADDITIONAL REFERENCES

Allee, W. C. 1931. *Animal aggregations: A study in general sociology.* Chicago: Univ. of Chicago Press.

Allee, W. C.; Allee, M. N.; et al. 1947. Leadership in a flock of white Pekin ducks. *Ecology* 28:310–15.

Cairns, R. B. 1966. Attachment behavior of mammals. *Psychological Review* 73:409–26.

———. 1966. Development, maintenance and extinction of social attachment behavior in sheep. *Journal of Comparative and Physiological Psychology* 62:298–306.

Cairns, R. B., and Werboff, J. 1967. Behavior development in the dog: An interspecific analysis. *Science* 158:1070–72.

Collias, N. E. 1952. The development of social behavior in birds. *Auk* 69:127–59.

———. 1956. The analysis of socialization in sheep and goats. *Ecology* 37:228–38.

Collias, N. E.; Collias, E. C.; Hunseker, D.; and Minning, L. 1966. Locality fixation, mobility, and social organization within an unconfined population of red jungle fowl. *Animal Behaviour* 14:550–59.

Enders, R. K. 1945. Induced changes in the breeding habits of foxes. *Sociometry* 8:53–55.

Fuller, J. L., and Fox, M. W. 1969. The behaviour of dogs. In *The behaviour of domestic animals,* ed. E. S. E. Hafez, pp. 438–81. 2d ed. London: Ballière, Tindall and Cassell.

Gray, P. H. 1963. The descriptive study of imprinting in birds from 1873 to 1953. *Journal of General Psychology* 68:333–46.

Hall, K. R. L., and Devore, I. 1965. Baboon social behavior. In *Primate behavior*, ed. I. Devore, pp. 53–110. New York: Holt, Rinehart & Winston.

Harlow, H. F., and Harlow, M. K. 1965. The affectional systems. In *Behavior of nonhuman primates*, ed. A. M. Schrier, H. Harlow, and F. Stollnitz, pp. 287–334. vol. 2. New York: Academic Press.

Murie, A. 1944. *The wolves of Mount McKinley.* U.S. Department of the Interior, Fauna Series, no. 5. Washington, D.C.: U.S. Government Printing Office.

Schutz, F. 1965. Sexuelle Prägung bei Anatiden. *Zeitschrift für Tierpsychologie* 22:50–103.

Scott, J. P. 1944. An experimental test of the theory that social behavior determines social organization. *Science* 99:42–43.

———. 1953. Implications of infra-human social behavior for problems of human relations. In *Group relations at the crossroads,* ed. M. Sherif and M. O. Wilson, chap. 2. New York: Harper & Bros.

———. 1956. The analysis of social organization in animals. *Ecology* 37:213–20.

———. 1969. Biological basis of human warfare: An interdisciplinary problem. In *Interdisciplinary relationships in the social sciences,* ed. M. Sherif, and C. W. Sherif, pp. 121–37. Chicago: Aldine.

Scott, J. W. 1942. Mating behavior of the sage grouse. *Auk* 59: 477–98.

Stewart, J. C., and Scott, J. P. 1947. Lack of correlation between leadership and dominance relationships in a herd of goats. *Journal of Comparative and Physiological Psychology* 40:255–64.

Tinbergen, N. 1953. *Social behaviour in animals.* London: Methuen.

Washburn, S. L., ed. 1961. *Social life of early man.* Viking Fund Publications in Anthropology no. 31. New York: Wenner-Gren Foundation.

Wheeler, W. M. 1923. *Social life among the insects.* New York: Harcourt-Brace.

Woolpy, J. H., and Ginsburg, B. E. 1967. Wolf socialization: A study of temperament in a wild social species. *American Zoologist* 7:357–63.

Young, S. P., and Goldman, E. A. 1944. *The wolves of North America.* Washington, D.C.: American Wildlife Institute.

Zajonc, R. B. 1968. Attitudinal effects of mere exposure. *Journal of Personality and Social Psychology* 9:1–27.

Chapter 9

RECOMMENDED GENERAL READING

Busnel, R. G., ed. 1963. *Acoustic behaviour of animals.* Amsterdam: Elsevier.
Good reviews by various authorities, organized according to major groups of animals.

Gould, J. E.; Henery, M.; and Macleod, M. C. 1970. Communication of direction by the honey bee. *Science* 169:544–54.
Experiments showing that bees can transmit information regarding the location of food without using odor clues.

Griffin, D. R. 1958. *Listening in the dark.* New Haven: Yale Univ. Press.
Detailed account of echolocation in bats, with some observation on hearing in other animals.

Hayes, C. 1951. *The ape in our house.* New York: Harper & Bros.
An account of Viki, one of the chimpanzees which have been brought up like human children.

Kellogg, W. N. 1961. *Porpoises and sonar.* Chicago: Univ. of Chicago Press.
Readable account of experiments on echolocation in these animals, with some information on their natural history and behavior.

Lanyon, W. E., and Tavolga, W. N., eds. 1960. *Animal sounds and communication.* Washington, D.C., American Institute of Biological Science.
Early symposium on auditory communication in animals.

Lilly, J. C. 1967. *The mind of the dolphin: A nonhuman intelligence.* Garden City, N.Y.: Doubleday.
Account of an experiment in trying to teach a dolphin human speech.

Lindauer, M. 1961. *Communication among social bees.* Cambridge: Harvard Univ. Press.
Continues account of work done by Von Frisch and associates, with comparisons of communication in bees other than the honey bee.

Sebeok, T. A., ed. 1968. *Animal communication.* Bloomington: Indiana Univ. Press.
Summary articles covering the major animal groups by a variety of authorities in the field.

Thorpe, W. H. 1961. *Bird song; The biology of vocal communication and expression in birds.* Cambridge: Cambridge Univ. Press.
Summary work by an authority in the field.

Von Frisch, K. 1967. *The dance language and orientation of bees.* Translated by L. E. Chadwick. Cambridge, Mass., Belknap Press of Harvard Univ. Press.
An excellent, well-written account of the author's studies, with many interesting facts about bees.

ADDITIONAL REFERENCES

Bronson, F. H. 1969. Pheromonal influences on mammalian reproduction. In *Perspectives in reproduction and sexual behavior,* ed. M. Diamond. Bloomington: Indiana Univ. Press.

Brower, L. 1969. Sex pheromones of the queen butterfly: Biology. *Science* 164:1170–72.

Bruce, H. M. 1960. A block to pregnancy in the mouse caused by the proximity of strange males. *Journal of Reproductive Fertility* 1:96.

Darwin, C. 1872; 1898. *The expression of the emotions in man and animals.* London: John Murray; New York: Appleton & Co.

Esch, H.; Esch, I.; and Kerr, W. E. 1965. Sound, an element common to communication of stingless bees and to dances of the honey bee. *Science* 149:320–21.

Frings, H., and Jumber, J. 1954. Preliminary studies on the use of specific sounds to repel starlings (*Sturnus vulgaris*) from objectionable roosts. *Science* 119:318–19.

Gardner, R. A., and Gardner, B. T. 1969. Teaching sign language to a chimpanzee. *Science* 165:664–72.

Hayes, K. J., and Hayes, C. 1951. The intellectual development of a home-raised chimpanzee. *Proceedings of the American Philosophical Society* 95:105–9.

Hinde, R. A., ed. 1969. *Bird vocalizations: Their relations to current problems in biology and psychology.* New York: Cambridge Univ. Press.

Kellogg, W. N. 1968. Communication and language in the home-raised chimpanzee. *Science* 162:423–27.

Mykytowycz, R. 1965. Further observations on the territorial function and histology of the submandibular cutaneous (chin) glands in the rabbit, *Oryctolagus cuniculus. Animal Behaviour* 13:400–412.

Nottebohm, F. 1970. Ontogeny of bird song. *Science* 167:950–56.

Schultze-Westrum, T. 1969. Social communication by chemical signals in flying phalangers (*Petaurus breviceps papuanus*). In *Olfaction and taste,* ed. C. Pfaffman, pp. 268–77. New York: Rockefeller Univ. Press.

Schusterman, R. J. 1966. Perception and determinants of underwater vocalization in the California sea lion. In *Les systèmes sonars animaux,* ed. R. G. Busnel, pp. 1–89. Frascati, Italy.

Weeden, J. S., and Falls, J. B. 1959. Differential responses of male oven birds to recorded songs of neighboring and more distant individuals. *Auk* 76:344–51.

Wenner, A. M., and Johnson, D. L. 1967. Honey bees: Do they use direction and distance information provided by their dancers? *Science* 158:1072–77.

Whitten, W. K. 1966. Pheromones and mammalian reproduction. In *Advances in reproductive physiology,* ed. A. McLaren, pp. 155–77. New York: Academic Press.

Whitten, W. K.; Bronson, F. H.; and Greenstein, J. A. 1968. Estrus-inducing pheromones of male mice: Transport by movement of air. *Science* 161:584–85.

Wilson, E. O. 1968. Chemical Systems. In *Animal communication,* ed. T. A. Sebeok, pp. 75–102. Bloomington: Indiana Univ. Press.

Yerkes, R. M., and Learned, B. W. 1925. *Chimpanzee intelligence and its vocal expression.* Baltimore: Williams & Wilkins.

Chapter 10

RECOMMENDED GENERAL READING

Allee, W. C. 1951. *Cooperation among animals.* Rev. ed. New York: Henry Schuman.
A popularly written account of the author's experiments on various aspects of social behavior and organization.

Allee, W. C.; Emerson, A. E.; Park, O.; Park, T.; and Schmidt, K. P. 1949. *Principles of animal ecology.* Philadelphia: W. B. Saunders.
Excellent detailed reference book on the organization of populations and social groups.

Blair, W. F. 1953. Population dynamics of rodents and other small mammals. *Advances in genetics,* 5:1–41. New York: Academic Press.
Review of evidence for territoriality and localization in these animals.

Calhoun, J. B. 1963. *The ecology and sociology of the Norway rat.* Bethesda, Md.: U.S. Dept. of Health, Education & Welfare.
A classical work on a captive population of wild Norway rats.

Esser, A. H., ed. 1971. *The use of space by animals and men.* New York: Plenum Press.
Papers and discussion by a group of world authorities.

Griffin, D. R. 1965. *Bird migration.* London: Heinemann.
Popularly written account of experiments done before this date.

Hediger, H. 1950. *Wild animals in captivity.* London: Butterworth.
On the application of the principles of animal behavior to the management of animals in zoological parks.

Howard, E. 1920; 1948. *Territory in bird life.* London: John Murray; London: Collins.
Presents the evidence for territoriality and its relation to bird song. The book was a major advance in the understanding of bird behavior.

King, J. A. 1955. *Social behavior, social organization, and population dynamics in a black-tailed prairie dog town in the Black Hills of South Dakota.* Contributions from the Laboratory of Vertebrate Biology, no. 67. Ann Arbor: Univ. of Michigan Press.
Monograph on the behavior under natural conditions of one of our most highly social rodents.

Klopfer, P. H. 1969. *Habitats and territories: A study of the use of space by animals.* New York: Basic Books.
An introduction to the problems of habitat selection and territoriality.

Matthews, G. V. T. 1968. *Bird navigation.* 3d ed. Cambridge, Cambridge Univ. Press.
Presents the argument and evidence for the "navigation theory" of bird orientation.

Mech, L. D. 1966. *The wolves of Isle Royale.* Fauna of the National Parks of the United States, Fauna Series 7. Washington: U.S. Government Printing Office.
Especially good on hunting behavior.

Nice, M. M. 1941. The role of territory in bird life. *American Midland Naturalist* 26:441–87.
Review of the history and supporting evidence for Howard's theory of territoriality, containing much additional information on the habits of American birds.

Russell, E. S. 1938. *The behaviour of animals.* 2d ed. London: Edward Arnold.
Written from the ecological viewpoint, this book emphasizes the behavior of invertebrates with many interesting examples.

Rutter, R. J., and Pimlott, D. H. 1968. *The world of the wolf.* Philadelphia: Lippincott.
An excellent and accurate popularly written account of wolf behavior; by two experts in the field.

Schaller, G. B. 1967. *The deer and the tiger.* Chicago: Univ. of Chicago Press.
Field studies of the population ecology and behavior of chital deer, certain other hoofed mammals, and tigers, in India.

Wynne-Edwards, V. C. 1962. *Animal dispersion in relation to social behavior.* Edinburgh: Oliver & Boyd. A comprehensive work dealing with the self-regulation of animal populations.

ADDITIONAL REFERENCES

Bronson, F. 1964. Agonistic behavior in woodchucks. *Animal Behaviour* 12:470–78.

Buechner, H. K. 1950. An evaluation of restocking with pen-reared bobwhite. *Journal of Wildlife Management* 14:363–77.

Calhoun, J. B. 1948. Mortality and movement of brown rats (*Rattus norvegicus*) in artificially supersaturated populations. *Journal of Wildlife Management* 12:167–72.

———. 1950. The study of wild animals under controlled conditions. *Annals of the New York Academy of Sciences* 51:1113–22.

———. 1952. The social aspects of population dynamics. *Journal of Mammalogy* 33:139–59.

Calhoun, J. B., and Webb, W. L. 1953. Induced emigrations among small mammals. *Science* 117:358–60.

Christian, J. J. 1963. Endocrine adaptive mechanisms and the physiologic regulation of population growth. In *Physiological mammalogy,* ed. W. V. Mayer and R. G. Van Gelder, pp. 189–353. New York: Academic Press.

Davis, D. E. 1966. The moult of woodchucks (*Marmota monax*). *Extrait de Mammalia* 30:640–44.

Eaton, R. L. 1970. Group interactions, spacing, and territoriality in cheetahs. *Zeitschrift für Tierpsychologie* 10(13):1–88.

Emlen, S. T. 1970. Celestial rotation: Its importance in the development of migratory orientation. *Science* 170:1198–1201.

Errington, P. L. 1963. *Muskrat populations.* Ames, Iowa: Iowa State Univ. Press.

Hasler, A. D. 1966. Underwater guideposts for migrating fishes. In *Animal orientation and navigation,* ed. R. M. Storm, pp. 1–21. Corvallis, Oregon: Oregon State Univ. Press.

Jacobs, M. E. 1955. Studies on territorialism and sexual selection in dragonflies. *Ecology* 36:566–86.

Lack, D. 1967. *The natural regulation of animal numbers.* 2d ed. Oxford: Oxford Univ. Press.

Lin, N. 1963. Territorial behavior in the Cicada killer wasp, *Sphecius speciosus* (Drury) (Hymenoptera; Sphecidae). *Behaviour* 20:115–33.

Pearl, R. 1932. The influence of density of population upon egg production in *Drosophila melanogaster. Journal of Experimental Zoology* 63:57–84.

Rabb, G. B.; Woolpy, J. H.; and Ginsburg, B. E. 1967. Social relations in a group of captive wolves. *American Zoologist* 7: 305–11.

Schmidt-Koenig, K. 1965. Current problems in bird orientation. In *Advances in the study of behavior,* ed. D. S. Lehrman, R. A. Hinde, and E. Shaw. New York: Academic Press.

Schneirla, T. C.; Brown, R. Z.; and Brown, F. C. 1954. The bivouac or temporary nest as an adaptive factor in certain terrestrial species of army ants. *Ecological Monographs* 24:269–96.

Schroeder, E. E. 1968. Aggressive behavior in *Rana clamitans. Journal of Herpetology* 1:95–96.

Southwick, C. H. 1955. Regulatory mechanisms of house mouse populations: Social behavior affecting litter survival. *Ecology* 36:627–34.

Terman, C. R. 1968. Population dynamics. In *Biology of Peromyscus* (*Rodentia*), ed. J. A. King, pp. 412–50. American Society of Mammalogists. Special publication no. 2.

Test, F. H. 1954. Social aggressiveness in an amphibian. *Science* 120:140–41.

Chapter 11

RECOMMENDED GENERAL READING

Dobzhansky, T. 1962. *Mankind evolving: The evolution of the human species.* New Haven: Yale Univ. Press.
Human evolution from the geneticist's viewpoint.

Etkin, W. 1967. *Social behavior from fish to man.* Chicago: Univ. of Chicago Press.
Review of evolution of social behavior in vertebrates, with a chapter by D. G. Freedman on a biological view of man's social behavior.

Hardin, G. 1959. *Nature and man's fate.* New York: Rinehart.
Chap. 12, "New dimensions of evolution" contains an excellent exposition of Wright's theories of evolution.

Hirsch, J., ed. 1967. *Behavior-genetic analysis.* New York: McGraw-Hill.
Technical reference work on methodology of relating genes to behavior, including a section on the evolution of behavior and its genetic basis.

Howells, W. W. 1959. *Mankind in the making: The story of human evolution.* Garden City, N.J.: Doubleday.
Human evolution from the viewpoint of a physical anthropologist.

King, J. A., ed. 1968. *Biology of Peromyscus (Rodentia).* Special Publication no. 2. The American Society of Mammalogists, Stillwater, Oklahoma.
General reference work on a genus containing 57 species occupying a variety of habitats and thus of special interest in the evolution of behavior.

ADDITIONAL REFERENCES

Baldwin, J. M. 1896. A new factor in evolution. *American Naturalist* 30:441–51, 536–53.

Cahalane, V. H. 1947. A deer-coyote episode. *Journal of Mammalogy* 28:36–39.

Emerson, A. E. 1943. Ecology, evolution and society. *American Naturalist* 77:97–118.

Firth, H. J. 1962. *The mallee-fowl: The bird that builds an incubator.* Sydney: Angus & Robertson.

Fisher, J., and Hinde, R. A. 1949. The opening of milk bottles by birds. *British Birds* 42:347–57.

Ford, E. B. 1964. *Ecological genetics.* New York: Wiley. Chap. 14, pp. 247–77.

Hamilton, W. B. 1964. Genetical evolution of social behavior. I and II. *Journal of Theoretical Biology* 7:1–52.

Hardy, G. H. 1908. Mendelian proportions in a mixed population. *Science* 28:49–50.

Harris, V. T. 1952. *An experimental study of habitat selection by prairie and forest races of the deermouse, Peromyscus maniculatus.* Contributions from the Laboratory of Vertebrate Biology, no. 56. Ann Arbor: Univ. of Michigan Press.

Huxley, J. S. 1923. Courtship activities of the red-throated diver (*Colymbus stellatus Pontopp*); together with a discussion of courtship behavior in grebes. *Journal of the Linnaean Society* 35:253–93.

King, J. A. 1961. Development and behavioral evolution in *Peromyscus.* In *Vertebrate speciation,* ed. W. F. Blair, pp. 122–47. Austin: Univ. of Texas Press.

Murie, A. 1940. *Ecology of the coyote in the Yellowstone.* U.S. Department of the Interior Fauna Series, no. 4. Washington, D.C.: U.S. Government Printing Office.

Roe, A., and Simpson, G. G., eds. 1958. *Behavior and evolution.* New Haven: Yale Univ. Press.

Scott, J. W. 1950. A study of the phylogenetic or comparative behavior of three species of grouse. *Annals of the New York Academy of Sciences* 51:1062–73.

Wecker, S. 1963. The role of early experience in habitat selection by the prairie deer mouse, *Peromyscus maniculatus bairdii. Ecological Monographs* 33:307–25.

Wright, S. 1940. Breeding structure of populations in relation to speciation. *American Naturalist* 74:232–48.

Wright, S. 1969. *Evolution and the genetics of populations.* Vol. 2. *The theory of gene frequencies.* Chicago: Univ. of Chicago Press.

Author Index

References to authors in illustrations are given in boldface type.

330 *Index*

Subject Index

Scientific names of particular animals mentioned in the text are listed in the index following the common name of the animals. For species of animals not listed in index, see under name of higher category, e.g., arthropods, vertebrates. References to drawings are printed in boldface type. Numbers in italics refer to the photographic inserts; the first number indicates the location of the insert, the second is the number of the page in the insert.

Accommodation
 related to principle that stimulus consists of change, 53
 sense organs, 7
Accuracy of experimenter as factor affecting experiments, 32
Adaptability, hereditary differences in, 132
Adaptation
 active and passive, 279–80
 effects of behavior, 9–10
 general systems of, 14–22, **16,** 60–65
 law of, 8–9
 and learning, 115
 and levels of organization, 270–76
 patterns of, not affected by environment, 120
 in relation to populations, 274–76
 in species and individual, 280
 theory of, xii
 see also Social adaptation
Adrenalin
 does not produce feeling of anger, 79
 effects on emotions, 79
African barkless dog, crossed with cocker spaniel, 126–30, **127,** *144–1*
Aggregations, 177–78
 and behavioral evolution, 282–83
 see also Shelter-seeking
Agonistic behavior, 73–82
 in baboons, 262 63
 of buffalo, *16–1*

definition of, 15–17
development of sex differences, 75
distinct from predation, 63
dog, *112–2*
effect of selection on, 296–97
evolution of, 284–87
functions of, 285–86
modification of, by selection, 123–24
and motor capacities, 62–63
of mouse, *112–2*
organized by hormones, 136–37
original stimulation from outside, 92
physiology of, 91
of sheep, 25
see also Escape behavior; Fighting behavior
Allelomimetic behavior
 in birds, *240–3*
 combined with aggressive fighting, 194
 definition, 18–19
 effect of sensory capacities on, 61–62
 of elk, *240–2*
 emotional basis, 89
 evolution of, 287–89
 of fish, *16–2*
 functions of, 287–89
 of geese, **20**
 and leadership, 189
 physiological mechanisms of, 89
 of sheep, 25
 and warfare, 194

333

344   *Index*

Octopus vulgaris
  discrimination learning in, 58–60,
    59, 114
  prehension in, 48–49
Odor. *See* Chemical senses
Operant conditioning apparatus, *112–*
  1
Order, as factor affecting behavior,
  31
Organization of behavior, 139–44,
  160–70
  capacities for, 57–60
  correlated with sensory and motor
    apparatus, 65
  in dogs, 155–57
  ecological, 236–46
  evolutionary changes in, 270–71
Orientation
  of bees, 228
  of birds, 237–40
Oven bird (*Seiurus aurocapillus*), response of to territorial calls, 216–
  17
Oxford, England, Bureau of Animal
  Populations, 248

Pain as primary stimulus in mouse
  fighting, 74
Paramecium caudatum
  attempts at adaptation of, 8
  avoiding reaction of, **6,** 8
Parasitism, limits behavioral variation, 281–82
Parental behavior. *See* Care-giving
  behavior
Passive inhibition, and control of aggression, 99, 107
Pavlovian apparatus, *112–1*
Peck order, 174–76, *176–1*
Penguin, Adelie (*Pygoscelis adeliae*)
  appendages of, *48–2*
  care-soliciting behavior, *240–4*
  colony of, *240–4*
  ecological barriers for, 246
  homing in, 238–39
Pheromones
  and communication, 230–32
  and hormones, 90
  primer, 232
Physical environment as factor affecting behavior, 31–32

Physiological counteraction and interaction, 80–82
Physiological mechanisms affecting
  fighting behavior, **81**
Pigeon (*Columba livia*)
  homing in, 237
  magnetic sense in, 240
Planaria (*Dugesia dorotocephala*),
  learning in, 113, 115, 293
Play as immature form of adult behavior, 26–27
Polarized light, reaction of bees and
  ants to, 229, 233
Polymorphism, and social organization, 284
Population
  control of, 268
  definition of, 246
  disorganized, 260–65
  of elk, 240–2
  and evolution, 276–78
  fluctuating, 247–51
  growth of, 257–66
  limits of, 258–60
  natural, 247–57
  natural selection of, 293
  produced by localization, 246
  social control of, 252–57
  stable, 252–57
Population dynamics, and history of
  science, xiv
Population genetics. *See* Genetics
Population problem in man, 268
Porpoise (*Tursiops truncatus*), and
  sonar, 220–21
Prairie dog (*Cynomis ludovicianis*),
  240–2
  social organization and stable pop-
    ulation of, 252
Preadaptation of behavior patterns,
  278–79
Predation
  defense against, 194
  distinct from agonistic behavior,
    63
  effects of on gulls, 151
Prehension, 48–51
Prenatal environment, selection pressure in, 295
Primary reactions, 154–57
  in ingestive behavior of dog, 96–97